CRUSADER
CASTLES

Carcassonne (August 1908)

CRUSADER CASTLES

BY
T. E. LAWRENCE

A New Edition
With Introduction and Notes
by
Denys Pringle

Clarendon Press · Oxford
1988

Oxford University Press, Walton Street, Oxford OX2 6DP

Oxford New York Toronto
Delhi Bombay Calcutta Madras Karachi
Petaling Jaya Singapore Hong Kong Tokyo
Nairobi Dar es Salaam Cape Town
Melbourne Auckland
and associated companies in
Berlin Ibadan

Oxford is a trade mark of Oxford University Press

Published in the United States
by Oxford University Press, New York

This edition © Oxford University Press 1988

British Library Cataloguing in Publication Data
Lawrence, T. E.
Crusader castles.—New ed.
1. Castles—Middle East   2. Architecture, Medieval—Middle East
I. Title II. Pringle, Denys
728.8'0956   NA497.M62/
ISBN 0-19-822964-X

Library of Congress Cataloging in Publication Data
Lawrence, T. E. (Thomas Edward), 1888–1935.
Crusader castles/by T. E. Lawrence.—New ed./with
introduction and notes by Denys Pringle.
1. Fortification—Middle East—History. 2. Fortification—Great
Britain—History. 3. Fortification—France—History. 4. Crusades.
5. Castles—Middle East—History. 6. Castles—Great Britain—
History. 7. Castles—France—History. I. Pringle, Denys. II. Title.
UG432.M628L38   1988   355.7'09—dc19   87-31390
ISBN 0-19-822964-X

Set by Wyvern Typesetting Ltd, Bristol
Printed in Great Britain
by Biddles Ltd., Guildford
and King's Lynn

FOREWORD

TO THE FIRST EDITION (1936)
BY A. W. LAWRENCE

IN 1910, T. E. Lawrence, then an undergraduate of Jesus College, took the Final Examination in History at Oxford. Availing himself of a new regulation which allowed candidates to present a Thesis as an additional part of the examination, he wrote the essay now printed for the first time. He had, in the ordinary way, chosen a Special Subject from a list of such prescribed by the Board and he was obliged to keep his Thesis within the limits of that subject, which was 'The First Three Crusades, 1095–1193'. Accordingly he defined the scope of his Thesis as 'The Influence of the Crusades on European Military Architecture to the end of the Twelfth Century'. The length was not to exceed 12,000 words.

In order to write the Thesis he had visited nearly all the important castles in England and Wales, France, Syria and northern Palestine. His European journeys were made on a bicycle (a three-speed machine with one unusually high gear, built to his order in a shop at Oxford by Mr Morris, who subsequently became famous in the motor industry). He rode all over England and Wales in a number of short trips, photographing and making plans of castles. In 1906 and 1907 he went to the north of France, traversing Normandy very thoroughly. In the summer vacation of 1908 he rode from Le Havre through the east of France to the Mediterranean, and back through the west, zigzagging through Les Andelys, Gisors, Coucy, Châtillon, Le Puy, Avignon, Les Baux, Aigues-Mortes, Carcassonne, Albi, Niort and Chartres. A first visit to the East followed in the summer vacation of 1909. He travelled on foot during the three hottest months of the year, with a passport obtained by Lord Curzon from the Turkish Government which secured him from official suspicion and gave him access to any buildings he wished to see. He landed at Beirut, went south to the Sea of Galilee, and then turned north to Aleppo: there he hired a carriage to take him to Urfa (Edessa). He returned to Beirut by Damascus.

The regulations required that a thesis should be sent to the examiners before the next Easter vacation so that he had no further opportunity for travel. Adequate descriptions of castles in other countries were rarely available, except for some in North Africa, so that he was unable as a rule to write of buildings he had not visited.

A first typewritten text of the Thesis served as a draft for a corrected version, also typewritten, which was bound with numbered illustrations and submitted to the examiners. The author then began to revise the rough copy again in pencil, apparently with a view to publication, altering the wording and writing comments in the margins. The title 'Crusader Castles' appears only in this version. After the examination (in which he was placed in the First Class), one of the judges of the Thesis urged the University Press to publish it in book form, saying, however, that not one of the illustrations could be spared; the cost was considered prohibitive. In later years the author added further notes in pencil, usually on the Examiners' Copy, which had been returned to him by special leave for the sake of the illustrations: he also placed a few more photos and plans inside this copy, some without reference to the existing text. But his occupations prevented him from visiting many more castles, and he was averse to publication until he had filled the gaps in his knowledge and could give a conclusive answer to the question implicit in the original title. Since most of the gaps still remain gaps in the world's knowledge and the question stands as before, I am now publishing a limited edition.

The text as printed incorporates the pencilled changes; it and the numbered footnotes, so far as they were presented to the examiners, are in Roman type. Italic type is used for pencil comments. These are not reproduced in every case; where notes to the same effect exist in both copies, only the fuller form is printed, and directions for more illustrations are omitted. So too are the Arabic transcriptions of place names, which were written within brackets in the text of the Examiners' Copy. To facilitate reading, section headings have been amplified in accordance with the manuscript Scheme [reproduced here in the Table of Contents on page xi]: in the original, numbers are sometimes used without headings and these numbers do not always correspond with those of the Scheme. Any other changes are enclosed in square brackets, except that apparent typist's errors have been corrected (on p. 74 'The design of the buildings of the hospital were . . .' is the original reading) and references are altered in conformity with new conditions. One duplicate illustration (62) inserted for the convenience of the examiners has been omitted, and the reference to it changed. The two little plans of El Mina and Provins, originally numbered 75 and 76 respectively, are jointly numbered 76 to allow of the inclusion as 75 of a better plan of Provins which was added to the Thesis after its return. The photographs I have called 11a and 50a were so added. A few extra illustrations are now included, selected from a quantity of material preserved separately. One of these is a rough sketch, made on the spot, here reproduced next to the drawing worked up at Oxford on that basis (23). The other additions are those numbered 2a, 74a, 78a, 91a, 92a, and the maps showing the network of castles visible from one another. These maps have necessarily been redrawn; the author traced them from originals by Mr H. Pirie-Gordon, DSC, FSA, who has kindly given much assistance with the reproduction in this edition.

. . .

A.W.L.

FOREWORD

TO THE PRESENT EDITION

THE text of *Crusader Castles* published in this edition follows that of the Examiners' Copy of the Thesis, which was published in the first volume of the Golden Cockerel edition of 1936. The material contained in the second volume of that edition, comprising letters written by T. E. Lawrence during his travels in France and the Levant, has since been fully published in *The Home Letters of T. E. Lawrence and his Brothers*, ed. M. R. Lawrence (Oxford, 1954). It is therefore not reprinted here.

In addition to the footnotes that formed part of the original Thesis, the 1936 edition also contained a selection of the marginal notes added later by Lawrence to the two typescript copies of the Thesis. Most of these were evidently intended to assist the author in his eventual revision of the work for publication. Here I have retained as footnotes only those which contribute new information or ideas to the argument, and have omitted the others, which are often no more than criticisms of style (e.g. 'Will make this stronger', 'Rot!', 'Quote examples', 'You've said this before'). Marginal notes taken from the Examiners' Copy (now in the Library of Jesus College, Oxford, MS J/160/7) are indicated in this edition by [X], those from the author's Rough Copy (in the Houghton Library, Harvard University) by [R]. References to ancient and medieval authorities have been updated to take account of more recent editions and, where possible, of English translations. Additional editorial notes are enclosed in square brackets.

The numbering of the illustrations accompanying the Thesis printed here (Figs. 1–96) follows that of the 1936 edition with the following changes. Additional illustrations include Figs. 2*b*, 11*b*, 13*a*, 17*a* (from Thesis), 21*a*, 33*a* (from Thesis), 39*a*, 43*a*, 46*a*, 50*b*, 64, 75*a*, 84*a*, 87*a*, 90*a* and 92*b*; and Pirie-Gordon's maps are omitted.

The present edition concludes with extracts from a preliminary draft of the Thesis, in which Lawrence's views on the relationships between Western, Byzantine and Crusading fortifications in the twelfth century are perhaps more clearly stated than in the more developed argument presented in the final version. This is followed by two extracts from Lawrence's later writings, in which he returned to the subject of Crusader castles. The first, contained in a letter of 1911, gives an account of the strategic siting of castles in the Crusader Levant. The second describes five castles in the County of Edessa, and was written during the course of an expedition, again on foot, in the same year. Lawrence's own photographs and drawings which accompany these extracts (Figs. 97–108) are numbered consecutively with those of *Crusader Castles*.

No attempt has been made to normalise the spellings of Arabic place-names referred to by Lawrence in these various writings, though Western ones are corrected where necessary. A key to the different names and spellings, however, will be found in the Index.

In preparing this edition I would like to record my thanks first of all to Professor A. W. Lawrence, who asked me to undertake the task and who has encouraged and assisted me throughout. I am also most grateful to the T. E. Lawrence Trustees, and in particular to their Secretary, Mr Michael V. Carey, for permitting me to make use of copyright material, and to Mr D. S. Porter, Senior Assistant Librarian of the Bodleian Library, Oxford University, for allowing me access to the papers deposited there. The Principal and Fellows of Jesus College, Oxford, kindly consented to allow me to consult and reproduce material from the Examiners' Copy of the Thesis, and I am especially grateful to Dr D. A. Rees, the College Archivist, for arranging for photographic copies to be made of the illustrations. Other illustrative material has been kindly made available by the Liddell Hart Centre for Military Archives, King's College, London, the President and Fellows of Magdalen College, Oxford, and Mr Malcolm Brown. Mr Jeremy Wilson, the author of a new biography of T. E. Lawrence, has generously provided me with much useful material on Lawrence himself and the writing of *Crusader Castles*.

Work on this edition has occupied me, intermittently, for several years, during which time I have moved home and work-place over three continents. To the British School of Archaeology in Jerusalem I am indebted for the opportunity, from 1979 to 1984, to gain a first-hand knowledge of Crusader monuments in Palestine, Transjordan, Syria and Cyprus. Between 1984 and 1985, Dumbarton Oaks, Washington DC, provided me with the time and research facilities necessary for attempting to encompass the bewildering range of material from different medieval cultures which *Crusader Castles* opens up. Finally, my resulting efforts have benefited from the critical eye of my present colleagues in the Inspectorate of Ancient Monuments of the Scottish Development Department.

R.D.P.

Edinburgh
8 November 1986

CONTENTS

ABBREVIATIONS

Lawrence's Abbreviations

Allcroft
: H. Allcroft, *Earthwork of England: Prehistoric, Roman, Saxon, Danish, Norman, and Mediaeval* (London, 1908).

Diehl
: C. Diehl, *L'Afrique byzantine: Histoire de la domination byzantine en Afrique (533–709)* (Paris, 1896).

Oman
: C. W. C. Oman, *A History of the Art of War: The Middle Ages from the Fourth to the Fourteenth Century* (London, 1898).

Rey
: E. G. Rey, *Étude sur les monuments de l'architecture militaire des croisés en Syrie* (Collection des Documents inédits sur l'Histoire de France, 1ère série, Histoire politique; Paris, 1871).

Viollet-le-Duc
: E. E. Viollet-le-Duc, *Military Architecture*, trans. M. Macdermott, 2nd edn. (Oxford and London, 1879); from *Essai sur l'architecture militaire au Moyen Âge* (extract from *Dictionnaire raisonné de l'architecture française du xıe au xvıe siècle*; Paris, 1854).

Abbreviations used in the Editor's Notes

Dictionnaire raisonné
: E. E. Viollet-le-Duc, *Dictionnaire raisonné de l'architecture française du xıe au xvıe siècle*, 10 vols. (Paris, 1867–8).

Home Letters
: *The Home Letters of T. E. Lawrence and his Brothers*, ed. M. R. Lawrence (Oxford, 1954).

Letters
: *The Letters of T. E. Lawrence*, ed. D. Garnett (London, 1938).

LIST OF FIGURES

Except where stated otherwise, the Figures are taken from the Examiners' Copy of the Thesis, by kind permission of the Principal and Fellows of Jesus College, Oxford. Duplicate prints of some of the Thesis photographs, however, also exist in the Liddell Hart Centre for Military Archives, King's College, London; in these instances the better of the two prints is represented here, and reference to the King's College photographs is noted by the formula 'also in King's College, London'.

Maps

INTRODUCTION

BY DENYS PRINGLE

THE period between the Norman conquest of England in 1066 and the English loss of Normandy in 1204 is often regarded as a crucial one for the development of techniques of fortification in northern and western Europe. Only a century and a half separates the crude earth and timber castles represented on the Bayeux Tapestry from King Richard I's Château Gaillard and the series of mighty stone-built fortresses with which King Philip Augustus of France secured his new conquests. The changes, however, affected more than just the materials employed in building castles. The earliest Norman stone castles such as Falaise (*c*.994) or the Tower of London (1077) were little different in conception from contemporary castles of wood and earth. A strong-point, be it a rectangular stone tower or a wooden structure raised on an earthen mound, constituted both the principal lordly residence and a place of refuge in time of attack. Surrounding it, or to one side, would be an outer enclosure, or bailey, more lightly defended and containing additional accommodation, stabling, stores and so on. The security of the place depended more on the solidity of its structures, especially that of the principal refuge, than on any particular attempt by its designers to allow the defenders to strike back advantageously at those attacking them.

Castles such as these, with a stone tower (keep or donjon) more often than not replacing the earlier and probably cheaper expedient of heaping up a large mound, continued to be built throughout the twelfth century and even later. But adequate though they might be as a lord's defence against his immediate neighbours or an unruly peasantry, their vulnerability to sustained mining or battering by a determined enemy became all the more apparent in the theatre of operations between the warring Plantagenets and Capetians and their respective vassals in the later part of the twelfth century. To contend with the threats posed by developing techniques of siege and sap, the plain rectangular type of tower-keep was sometimes modified by adding extra buttressing to the corners and sides; its base could also be thickened with a smooth masonry batter or talus; and corbelled *machicoulis* might also be constructed to allow the defenders on the wall-top to rain down missiles on anyone attempting to approach its base. Experiments were also made in altering the shape of the keep, so as to make it better able to withstand the impact of an enemy's stone-hurling artillery. Some, for instance, were provided with a pointed prow or beak, facing the most exposed quarter and intended to deflect missiles. And polygonal or circular keeps, with or without buttressing, also made their appearance in the later twelfth and thirteenth centuries. Finally the keep could be dispensed with altogether; or, put another way, the whole castle could become in effect a keep. The straggling trace of earlier bailey walls would thus be replaced by smaller, more compact enceintes, either following the contours of the site or, where this lay in open flat terrain, conforming to a regular geometrical ground-plan. In either case, no part of the enceinte would be allowed to be weaker than any other. The walls would be provided with rows of casemated arrow-slits, reached from continuous walkways (or *chemins de ronde*) set inside the wall, often at more than one level; and each section of curtain would be overlooked by one or more adjacent towers. Elaborate gateways with combinations of wing-doors, portcullises and carefully sited arrow-slits and machicolations lessened the chance of the garrison being taken by surprise.

In this way, the 'passive' resistance of earlier castles, in which the defenders were forced to retreat into progressively more strongly fortified parts of the castle, was replaced by a system designed to give them the opportunity to play a more 'active' role in attacking the enemy before he could ever breach the outer defences. In addition to a ditch, these outworks might now include an outer wall, also provided with towers, but of slighter proportions than the inner one, so that the enemy could be engaged by the defenders of both walls at once and so that, in the event of the outer wall being taken, it would give small advantage to the attackers, who would still find themselves overlooked by the towers and curtains of the inner wall.[1]

Expressed in such terms, this scheme of development represents, of course, an over-simplification and conceals the great diversity of castle types constructed throughout north-western Europe in the eleventh and twelfth centuries. In general it relates only to the more 'advanced' types of castle being built; and even among these there are a number which deviate from the general model. Framlingham, for instance, represents a late eleventh- to twelfth-century castle of stone which never had a keep;[2] while at the other extreme it has been noted that

---

[1] For developments in French and English castle design to the end of the 13th century, see the general treatments by J.-F. Finò, *Forteresses de la France médiévale* (Paris, 1970), 95–108, 156–79, 215–27; G. Fournier, *Le Château dans la France médiévale: Essai de sociologie monumentale* (Paris, 1978), 65–99; R. A. Brown, *English Castles* (London, 1976), 40–127.

[2] Brown, *English Castles*, p. 65, fig. 26.

rectangular keeps continued to be built in Angevin England and Normandy long after the type had become outmoded in the rest of France.[3] Nevertheless, when viewed over the space of two centuries, from around 1100 to around 1300, such a scheme of development may be seen to have a general validity. Château Gaillard, built in the closing years of the twelfth century, represents a significant landmark on the road towards the keepless 'concentric' castle (see below, pp. 112–22). It still retains a keep, of novel design, but the main strength of the castle, apart from that conferred on it by its location on a natural spur overlooking the River Seine, lay in the skilful design of the outer walls, strengthened with projecting rounded towers so positioned that no part of it was without flanking fire. A certain amount of added support could also have been lent the defenders of the outer enceinte by those of the inner, and even perhaps the keep, firing over their heads. Had Château Gaillard been constructed in a plain instead of on a hill, it might more readily be hailed by historians and archaeologists as the earliest consistent attempt at 'concentric' castle-planning in north-western Europe. The finest expression of the type, however (before, that is, the advent of gunpowder brought about a complete revolution in techniques of fortification in the late fifteenth century), may be seen in the castles with which King Edward I and his Savoyard architect, Master James of St George, secured the English conquest of Wales at the close of the thirteenth century.[4] At Beaumaris Castle in Anglesey (1295–) there is no keep at all. The defences consist of two roughly rectangular enclosures of stone, one inside the other, the whole castle being surrounded by a water-filled moat. Both walls are strengthened by rounded towers, the inner wall and its towers overtopping the outer; and the approaches to the gates run obliquely through the lists (the space between the two walls) exposing anyone approaching the inner gate-houses to flanking fire from one or both of the walls.[5]

The science with which such castles as Beaumaris, Harlech and Rhuddlan were fortified, however, was not an entirely new development of the later thirteenth century. Finer details of architecture and construction apart, these works may be considered as representing no more than a reapplication of principles for defending fortified places that had been known, and on occasion practised, since Hellenistic times and had been committed to posterity in the writings of men such as

[3] P. Héliot, 'L'évolution du donjon dans le nord-ouest de la France et en Angleterre au xiie siècle', *Bulletin archéologique du Comité des Travaux historiques*, NS 5 (1969), 141–94 (pp. 156–7).

[4] Brown, *English Castles*, pp. 95–111; A. J. Taylor, in H. M. Colvin (ed.), *History of the King's Works*, i–ii. *The Middle Ages* (HMSO; London, 1963), i. 293–408 (repr. as *The King's Works in Wales, 1277–1330* (HMSO; London, 1974)).

[5] Taylor, *History of the King's Works*, i. 395–408; id., *Beaumaris Castle, Castell Biwmares, Gwynedd* (HMSO; Cardiff, 1980).

Philo of Byzantium in the fourth century BC.[6] To historians of the later nineteenth century, accustomed to regard advances in Western culture in the later Middle Ages in terms of a progressive rediscovery of the classical excellence of ancient Greece and Rome which had been so thoughtlessly destroyed by the barbarians in the fifth century, it was quite natural to seek an explanation for the advances discernible in European castle-building in terms of borrowings and relearnings from that earlier, higher source of civilisation. Medieval Byzantium, for all its apparent failings, represented an obvious link between the ancient and medieval worlds; and indeed, Greek theories of fortification had been kept alive and elaborated further in Byzantine military treatises written in the sixth century AD and later. From 1097, for more than two centuries, the Crusading movement brought Western kings, princes, magnates and their vassals face to face with the Byzantine world of the eastern Mediterranean and provided an obvious context in which the adoption and transmission of ideas could take place. And the likelihood that ideas of fortification did pass from East to West was given added weight by the consideration that the two builders of the most advanced systems of fortification in north-western Europe at the end of the twelfth century, Kings Richard I of England and Philip Augustus of France, had both participated in the Third Crusade to the Holy Land (1189–92), where they would doubtless have seen the novel types of town and castle defence that the Franks of Outremer were already adopting under Byzantine influence.

The argument was indeed seductive. Its principal strength and weakness was that so little work had been done to record systematically the works of medieval fortification in the West, let alone in the East, that it was virtually impossible either to prove or disprove it. And few of the scholars who wrote with such apparent authority of the importance of Crusader castles for the development of fortification techniques in Europe had ever been to the Levant, let alone seen a Crusader castle.

Before going up to Oxford in 1907, T. E. Lawrence was already familiar with a large number of castles in England, Wales and France through cycling tours made with his father and later with friends. His knowledge of the castles of France was strengthened by a further trip made during the summer vacation following his first year at the University, which took him to the Mediterranean and back. Letters written on that

[6] Among recent surveys of Hellenistic fortifications may be noted A. W. Lawrence, *Greek Aims in Fortification* (Oxford, 1979), which contains an annotated translation of the relevant passages from Philo's treatise (pp. 73–107). See also Y. Garlan, *Recherches de poliorcétique grecque* (Bibl. des Écoles françaises d'Athènes et de Rome, 223; Paris, 1974).

occasion show that he was already considering writing a thesis on castles as part of his final examination in History. It was the chance of a three-and-a-half-month visit to Syria and northern Palestine the following summer, however, that enabled him to extend the scope of his enquiry and consider precisely that question which so many previous commentators on European castles had taken for granted, 'The Influence of the Crusades on European Military Architecture to the end of the Twelfth Century'.

Lawrence outlines his approach to this question in his introductory chapters (pp. 3–5). Only by making a direct comparison between the development of fortifications in Western Europe and that in the Eastern Empire up to the beginning of the twelfth century would it be possible to establish how far the architects of Crusader Syria had been indebted to either source; and by comparing their castles directly with those being erected in Europe in the same period, the contrast between them would become obvious.

The first castles occupied by the Crusaders in northern Syria and in what is today south-eastern Turkey were, as Lawrence explains, often no more than sites fortified at an earlier date by the Byzantine Greeks. But because many of them were also refortified in later periods, it was hard to distinguish the characteristics that might have been particular to whatever the Crusaders may have added to them. Further south, however, in the County of Tripoli and the Kingdom of Jerusalem itself, Lawrence found that the earliest type of fortification, built *de novo* or added to existing works by the Crusaders, was in many cases a rectangular keep or donjon. Although differing in some respects from European keeps, for example by having stone in place of wooden floors, no more than two storeys and an entrance normally at ground- as opposed to first-floor level, there was no doubt that here at least was a type of fortification which the Franks had brought with them from the West.

The period of keep-building was followed, mid-century, by one which Lawrence describes as characterised by the 'clash of home and Greek styles', out of which two 'rival' styles developed. One of these, championed by the Templars, followed the Greek pattern, as Lawrence understood it, of slight single curtain walls, weak rectangular towers of small projection and the absence of hoarding or machicolation. The other, adopted by the Hospitallers, was French in influence and was characterised by the use of rounded towers, a masonry talus, various forms of machicolation and 'concentric' planning. He concludes, 'The only people wholly independent of Europe had been the Templars, and their style was practised only by themselves, and died with them. All the best of the Latin fortifications of the Middle Ages in the East was informed with the spirit of the architects of central and southern France' (p. 88).

Turning to the castles being built in France during the same period, Lawrence presents a picture of progressive development, beginning with modifications made in the design of keeps and ending in the masterpiece of Château Gaillard. While not denying the possibility that certain minor architectural details of some French castles may have been the result of Eastern influence, Lawrence argued that the example of Crusader castles exerted no significant influence on the course of this development. The idea that two lines of defence were stronger than one would, he reasonably pointed out, have been a simple matter for the builders of early gothic cathedrals to figure out; 'multiple' or 'concentric' defences can be seen even in prehistoric earthworks; the classic work on fortification by Vegetius (and one could add Vitruvius) was available to and apparently even consulted by castle-builders; and machicolations were being constructed in France earlier than in Syria. Thus, he maintained, Château Gaillard was 'northern French in design, and north French in execution' (p. 112); and, he concluded, 'There is not a trace of anything Byzantine in the ordinary French castle, or in any English one: while there are evident signs that all that was good in Crusading architecture hailed from France or Italy' (pp. 116–22).[7]

Lawrence's Thesis was not published until 1936, a year after his death, and then only in a limited edition for private distribution. It was not apparently known by Paul Deschamps when he published his magisterial account of Crac des Chevaliers in 1934, although he was able to refer to it in the later volumes of his survey of Crusader castles.[8] Had Lawrence been able to prepare the text for publication himself, there is no doubt, as his marginal notes indicate, that he would have made extensive alterations and additions to the text. One such note (p. 83 n. 82), for example, indicates that the publication of Miss Gertrude Bell's monograph on Ukhayḍir, where buttress-machicolations of the type found at Crac were recorded on an eighth-century Muslim palace building, would have necessitated a revision of the idea that the Crusaders introduced the idea of *machicoulis* to the East. Nevertheless, his case that Crusader castles owed more to the West than to Byzan-

[7] The inclusion of Italy may have been due to the supposed similarity of certain machicolations at Qalʿat Ṣubayba to Provençal or Italian types suggested by W. R. Lethaby (p. 74). Elsewhere, however, Lawrence characterises Italian fortifications as showing a 'distinct Byzantine feeling' (p. 5), though he admits in a marginal note that he had never actually been to Italy.

[8] *Les Châteaux des croisés en Terre Sainte*, i. *Le Crac des Chevaliers* (Bibliothèque archéologique et historique, 19; Paris, 1934); ii. *La Défense du royaume latin de Jérusalem* (Bibl. archéol. et hist., 34; Paris, 1939); iii. *La Défense du Comté de Tripoli et de la Principauté d'Antioche* (Bibl. archéol. et hist., 90; Paris, 1973).

tium, and that they contributed little or nothing to the development of European castles before the end of the twelfth century, represented a significant contribution to a debate that has continued intermittently to this day.[9] Three-quarters of a century of research since Lawrence completed his Thesis has added considerably to our knowledge and understanding of castles in both East and West. Some of the results of this research, in so far as it affects the interpretation of individual sites, are indicated in the editorial notes which accompany Lawrence's text printed below. At this point, however, it may be appropriate to assess how far Lawrence's general conclusions remain valid and in what respects subsequent research makes it necessary for them to be modified.

It is often the case when attempting to answer questions of historical interpretation that as more information is gathered together so the original question itself has to be changed. The question to which Lawrence, and, before him, Sir Charles Oman had addressed themselves and had supplied conflicting answers concerned the amount of Greek influence to be detected in Frankish castles built in the East and in the West at the time of the Crusades. The assumption of both seems to have been that the local models that might have inspired Crusader builders were Byzantine. They therefore overlooked other more plausible sources of inspiration.

The possibility that Muslim architecture might have provided models for imitation is summarily dismissed by Lawrence (see pp. 29 and 35). Even where he acknowledges that the native workforce used by the Franks may have contributed at least some minor details of construction, his identification of these people as 'Syrians accustomed to build Greek fortresses' (p. 49) quickly removes from consideration the possibility that they would have had a cultural identity of their own. The walls of the coastal towns of Syria and Palestine which the Crusaders besieged and took in the first half of the twelfth century, however, had been built not by Greeks but by the Abbasids, Seljuks and

---

[9] Lawrence's Thesis has been commented upon by R. Fedden, *Crusader Castles* (London, 1950), 22–31; R. C. Smail, *Crusading Warfare (1097–1193)* (Cambridge Studies in Medieval Life and Thought, NS 3; Cambridge, 1956), 226, 237–8, 244 n. 1; R. Fedden and J. Thomson, *Crusader Castles* (London, 1957, repr. 1968), 41, 84; T. S. R. Boase, *Castles and Churches of the Crusading Kingdom* (Oxford, 1967), 105–8; R. C. Smail, *The Crusaders in Syria and the Holy Land* (Ancient Peoples and Places Series, 82; London, 1973), 94–6; T. S. R. Boase, 'Military Architecture in the Crusader States in Palestine and Syria', in H. W. Hazard (ed.), *The Art and Architecture of the Crusader States* (K. M. Setton (ed.), *A History of the Crusades*, 4; Madison, Wis., 1977), 140–64 (p. 140 n.). An exhaustive analysis of the work has also been made by M. Larès, *T. E. Lawrence, la France et les Français*, 2 vols. (Lille, 1978), i. 193–271; and *Lawrence d'Arabie et les châteaux des croisés* (Publns. de l'Association de médiévistes anglicistes de l'enseignement supérieur, 6; Paris, 1980).

Fatimids; and the walls of Jerusalem itself had been rebuilt by the city's Seljuk governor within sixty years of its falling to the army of the First Crusade in 1099. Little is known archaeologically of these works of fortification. But in Cairo, the walls and gates built by the Fatimids in the 1090s still survive. Here we find gateways built of fine ashlar, flanked by rounded towers and protected by *machicoulis*; and inside the curtain walls a covered *chemin de ronde* provided a secondary firing gallery below the crenellated wall-top.[10]

The architects responsible for the walls of Cairo are identified by the historian al-Maqrīzī as three Armenians from Edessa. The Kingdom of Armenia, wedged between Byzantine Asia Minor and the advancing Seljuk Turks, also seems likely to have been a source from which the builders of Crusader castles would have drawn inspiration in the twelfth and thirteenth centuries. That Armenian masons were working in the Crusader states as far south as the Kingdom of Jerusalem is indicated by their masonry marks, inscribed on ashlars of the Church of the Annunciation in Nazareth;[11] while the Armenian Cathedral of St James in Jerusalem, rebuilt by *c.* 1165, represents a blending of Western and Armenian styles of ecclesiastical architecture in a single building.[12] Armenians and Muslims alike were constructing castles with rounded or cylindrical towers several centuries before these became common in medieval Western Europe.[13]

A second difficulty besetting Lawrence's attempt to identify, or deny, Byzantine influence in Frankish castles was that his characterisation of Byzantine military architecture was based, of necessity, almost exclusively on Charles Diehl's treatment of the fortifications erected by Justinian and his immediate successors in North Africa in the sixth century.[14] These may have little relevance, however, for determining the influence that surviving Byzantine fortifications or contemporary Byzantine military practice could have had on Crusader fortifications of the twelfth century. Indeed, the works built in Africa in the sixth century do not even compare with the system of triple defences erected at Constantinople a century earlier; and these walls, unlike the African

---

[10] K. A. C. Creswell, 'Fortification in Islam before A.D. 1250', *Proceedings of the British Academy*, 38 (1952), 89–125 (pp. 112–19); id., *The Muslim Architecture of Egypt*, 2 vols. (Oxford, 1952–9), i. 161–217.

[11] B. Bagatti, *Gli Scavi di Nazaret*, ii. *Dal secolo XII ad oggi* (Studium Biblicum Franciscanum, Collectio maior, 17; Jerusalem, 1984), fig. 32, nos. 22, 23; cf. D. Pringle, 'Some Approaches to the Study of Crusader Masonry Marks in Palestine', *Levant*, 13 (1981), 173–99 (p. 177 n. 7).

[12] H. Vincent and F. M. Abel, *Jérusalem: Recherches de topographie, d'archéologie et d'histoire*, ii. *Jérusalem nouvelle*, 4 fascs. with album (Paris, 1914–26), 522–6, 530–58, figs. 197–225, pls. LIV–LVIII.

[13] Cf. H.-P. Eydoux, *Les Châteaux du soleil: Forteresses et guerres des croisés* (Paris, 1982), 334.

[14] C. Diehl, *L'Afrique byzantine: Histoire de la domination byzantine en Afrique (533–709)* (Paris, 1896), 138–225. The same criticism may also be made of Deschamps, *Châteaux*, i. 47–53.

ones, would have been seen by the leaders of the First Crusade and by countless other Westerners before and after. Neither do they compare favourably with sixth-century fortifications on the eastern frontier, such as Ruṣāfa (Sergiopolis), where a two-tiered *chemin de ronde*, a variety of differently shaped towers and an outer ditch and *proteichisma* (outer wall) represent a more sophisticated response to the serious threat posed by the siege techniques of the Persians.[15]

Secondly, Diehl's synthetic style of writing, in which archaeological and historical material is selected and welded into a single coherent line of argument, often fails to distinguish between what was usual and what was unusual in Byzantine military practice. Thus the *pyrgokastellon*, which he indicates as a forerunner of the medieval keep or donjon, is in fact mentioned only once in ancient sources, by Procopius describing Justinian's addition of extra towers (in the plural) to the walls of Constantina in Mesopotamia; far from being normal, these towers were exceptional, hence the unusual name coined for them (from *pyrgos*, Greek for 'tower', and *castellum*, Latin for 'fort'). Similarly, sixth-century towers were not normally isolated from the curtain walls and they did not normally have internal staircases; Procopius describes towers of this kind on the Long Wall in Thrace apparently because they departed from normal practice.[16]

The extent of direct borrowing made by the Crusaders from existing or contemporary Byzantine models can only properly be assessed after more detailed study has been made of those sites in northern Syria and south-east Turkey where Byzantine fortifications of the period immediately preceding the First Crusade exist. One site which falls into this category and which Lawrence visited in 1909 is Ṣahyūn. In the tenth century the eastern side of this promontory site was detached from the higher ground to the east by a triple line of walls, the inner one strengthened by cut-water-shaped towers, the outer two with cylindrical towers, before which there was a deep rock-cut chasm. As R. C. Smail has shown, here at least the Franks can be seen to have dismissed the earlier system as inadequate and constructed instead a massive wall strengthened by large towers, including one which may be described as a keep, on the line of the outer Byzantine wall; the Crusaders also

[15] On Ruṣāfa/Sergiopolis, see W. Karnapp, 'Die Stadtmauer von Resafe in Syrien: Vorläufiger Bericht', *Archäologischer Anzeiger* (1968), 307–43; 'Die Nordtoranlage der Stadtmauer von Resafa in Syria', *Archäologischer Anzeiger* (1970), 98–128; *Die Stadtmauer von Resafa in Syrien* (Denkmaler Antiker Architektur, 2; Berlin, 1976).

[16] For further discussion, see D. Pringle, *The Defence of Byzantine Africa from Justinian to the Arab Conquest*, 2 vols. (British Archaeological Reports, International Series, 99; Oxford, 1981), i. 155, 170. The danger of taking Procopius' descriptions of town defences at face value is highlighted by a recent archaeological and literary study: B. Croke and J. Crow, 'Procopius and Dara', *Journal of Roman Studies*, 73 (1983), 143–59.

appear to have deepened the outer ditch.[17] Although some Byzantine influence may be recognised in the two-tier *chemin de ronde* in the Frankish wall, the bent entrances of the two postern gates and the simple design of the arrow-slits, there is nothing specifically Byzantine about the design of these towers nor in their isolation from the *chemin de ronde* of the adjacent curtain walls.

Apart from Byzantine, Islamic and Armenian influences, other factors may also be expected to have played a part in the development of Crusader systems of fortification. The general use of stone vaulting, for instance, may well, as Lawrence suggests, have been the result of a lack of suitable timber; but the existence in most areas of the Levant of a ready supply of building stone and of masons experienced in working it, together with the common use made by Muslims and Greeks of incendiary devices in attacking fortified places, may have been factors no less significant.

Lawrence's most original contribution to the study of Crusader castles has been identified in his recognition that the earliest of them, built with a keep, conformed to a purely Western type of castle design.[18] Recent research confirms this. Indeed, it appears that in the area around Jerusalem the conquest was followed by the construction of large numbers of small keep-and-bailey castles, representing in most cases centres of newly established seigneurial domains.[19] Even some of the larger castles of the Crusader states, such as Beaufort (p. 58) and Tripoli,[20] seem originally to have been of this type. The construction of keeps, however, did not end in 1150. The *église-donjon* at Ṣāfīthā is now dated after 1170 (cf. pp. 55–8). And from 1226 onward, the Teutonic Knights built at Montfort Castle a vast D-shaped keep, to reinforce that part of the site most exposed to enemy attack.[21] Trapezoidal keeps were used in a similar fashion by the Muslims to strengthen the castles of Shayzar (1233) and Karak in Moab in the same period.[22]

---

[17] Smail, *Crusading Warfare*, pp. 236–9. On the Byzantine defences of the site, see A. W. Lawrence, 'A Skeletal History of Byzantine Fortification', *Annual of the British School at Athens*, 78 (1983), 171–227 (pp. 218–19, fig. 21, pl. 19*b*); Deschamps, *Châteaux*, iii. 217–47, plans, pls. VIII–XXX. [18] Smail, *Crusading Warfare*, p. 226.
[19] D. Pringle, *The Red Tower (al-Burj al-Ahmar): Settlement in the Plain of Sharon at the Time of the Crusaders and Mamluks (A.D. 1099–1516)* (British School of Archaeology in Jerusalem, Monograph Series, 1; London, 1986), 15–18.
[20] On Tripoli, see W. Müller-Wiener, *Castles of the Crusaders* (London, 1966), 42–3, plan 1, no. 2.
[21] B. Dean, *A Crusaders' Fortress in Palestine: A Report of Excavations made by the Museum, 1926* (Part II of the Bulletin of the Metropolitan Museum of Art; New York, 1927, repr. Jerusalem, 1982); M. Benvenisti, 'Montfort', in M. Avi-Yonah and E. Stern (eds.), *Encyclopedia of Archaeological Excavations in the Holy Land* (Oxford, 1977), iii. 886–8; id., *The Crusaders in the Holy Land* (Jerusalem, 1970), 331–7; R. D. Pringle, 'A Thirteenth-Century Hall at Montfort Castle in Western Galilee', *Antiquaries Journal*, 66 (1986), 53–81 (p. 54, fig. 3).
[22] Shayzar: Müller-Wiener, *Castles*, pp. 55–6, pl. 48; M. Van Berchem and E. Fatio, *Voyage*

The idea that there developed later in the twelfth and in the thirteenth century two 'rival' schools of Crusader castle-builders, a Hospitaller one inspired from France and a Templar one inspired by Byzantium, was first suggested by E. G. Rey.[23] In elaborating the same theory, Lawrence betrays like Rey an obvious delight for anything reminiscent of France and contempt for things supposedly Byzantine. The hypothesis was based, however, on a very small selection of known sites. When a broader range of castles founded by the two Orders is considered, it can no longer be sustained.[24]

It is clear, none the less, that the walls of the Templar castle of ʿAtlīt (Pilgrims' Castle) do indeed betray the influence, direct or indirect, of Hellenistic principles of fortification. As Smail has remarked, however, Lawrence's assessment of this castle (p. 71) makes strange reading in the light of the survey and excavations conducted by C. N. Johns in the 1930s.[25] These have revealed that the castle's landward defences consisted of a massive wall, 12 m. thick and over 30 m. high, flanked by two rectangular towers, each 27 × 21 m. overall and projecting between 6 and 16 m. from the wall face. Outside this and separated from it by a space of 27 m. was another, somewhat slighter wall, 6.5 m. thick and 16 m. high, with three rectangular towers so positioned that the archers and ballista-operators on the inner towers could fire between them and over the outer wall. Postern gates were set in the side walls of the towers. Finally, an outer rock-cut ditch with a vertical counterscarp wall completed the defences. The entrance through this system of fortification took a snaking path between the walls, leading the attackers directly under the fire of the defenders on the towers and curtains. The system of defence described thus far, scale and constructional techniques apart, was of a type which would have been familiar to Philo of Byzantium a millennium and a half earlier. But in addition, the curtains and towers of ʿAtlīt were provided with timber hoarding (as the evidence of putlog holes indicates), stone box-*machicoulis* and casemated arrow-slits with sloping sills, enabling the defenders to prevent the enemy approaching the base of the walls.[26]

The defences of the castle of Ṭarṭūs (see pp. 49, 73) are very similar to those of ʿAtlīt, but more extensive since here the sea provided protection on only one side instead of three. At Ṭarṭūs we find the same system of massive inner and outer walls, strengthened by projecting

en Syrie, 2 vols. (Mémoires de l'Institut français d'archéologie orientale du Caire, 37–8; Cairo, 1914–15), i. 177–88, pls. 25–7. Karak: Müller-Wiener, *Castles*, pp. 47–8, pls. 24, 26; Deschamps, *Châteaux*, ii. 88–9, pls. XI, XIX.

[23] *Les Colonies franques de Syrie aux xiime et xiiime siècles* (Paris, 1883), 119–37.

[24] Cf. Boase, 'Military Architecture', pp. 156–7; Smail, *Crusaders*, pp. 114–15.

[25] *Crusading Warfare*, p. 244 n. 1.   [26] For notes on ʿAtlīt see below p. 71.

towers so disposed that an attacking force could be engaged by the ballistae and archers of both walls at once, and surrounded by an outer rock-cut ditch. The main gate on the north led through a bent entrance within one of the towers and then along the lists to the west to reach the inner gateway. ʿAtlīt Castle never fell: it was unsuccessfully attacked by Sultan Baybars in 1265, and in 1291 its Templar garrison quietly abandoned it after the fall of Acre. Ṭarṭūs was the last Crusader stronghold on the coast of Syria to be evacuated by the Franks, in 1292.[27]

ʿAtlīt and Ṭarṭūs were both built in the first half of the thirteenth century. It is harder to detect such regularity in the planning of twelfth-century Templar castles, partly because the castles belonging to the Order in that period were not always ones that they had built them-selves, but also because many of them are now so badly ruined. The castle of La Fève (al-Fūla), however, which existed by 1171, appears to have been roughly rectangular in shape, probably with rectangular corner- and interval-towers and possibly with an inner and outer enceinte.[28] Latrūn (le Toron des Chevaliers) also seems to have had a rectilinear plan with rectangular towers; it was also probably 'con-centric', though the first castle was of the keep-and-bailey type.[29] ʿArayma (pp. 71–3) includes twelfth- and thirteenth-century work; it too consists of a rectangular inner ward with rectangular corner-towers, surrounded by an outer enceinte with a gateway flanked by a pair of rounded towers on the south.

Not all Templar castles in Syria and Palestine, however, had rec-tangular plans and rectangular towers. Ṣāfīthā (p. 59) had an elliptical enceinte, following the contours of its site, with a rectangular donjon added after 1170. At Ṣafad (pp. 66–8) the twelfth-century castle plan is uncertain. The thirteenth-century castle, however, built between 1240 and 1260, had a double enceinte, the outer with rounded towers and the inner with rectangular ones, with possibly a polygonal or cylindri-cal église-donjon at the highest point of the site.[30] Thus rectangular and 'concentric' planning are to be found in Templar castles of the twelfth century, and were to be more fully developed in those of the thirteenth. But there is no reason to believe that such systems were 'unintelligent' or weak; on the contrary, these castles included some of the most formidable ones with which the Mamluks ever had to contend. Neither

[27] For notes on Ṭarṭūs, see below pp. 49–51.

[28] B. Z. Kedar and D. Pringle, 'La Fève: A Crusader Castle in the Jezreel Valley', *Israel Exploration Journal*, 35 (1985), 164–79, pls. 20–1.

[29] M. Ben-Dov, 'The Fortress at Latrun', *Qadmoniot*, 7 (1974), 117–20 (for plan) [Hebrew]; cf. Benvenisti, *Crusaders*, pp. 316–18.

[30] D. Pringle, 'Reconstructing the Castle of Safad', *Palestine Exploration Quarterly*, 117 (1985), 139–49.

is French influence in their architecture entirely lacking, as the exquisite early gothic polygonal chapel and the suburban church at 'Atlīt illustrate.[31]

The features by which Rey and Lawrence sought to distinguish Hospitaller castles from those of the Templars also now appear to be mostly illusory. The rounded towers with machicolations at Qal'at Ṣubayba, for instance, are now known to be Ayyubid of the late twelfth or early thirteenth century. Indeed, it seems unlikely that any part of this castle was built by the Hospitallers (pp. 73–4). The defensive schemes at Crac and Marqab, that were held to typify the work of this 'school', date from the last quarter of the twelfth century and were continued in the thirteenth. Here we do find rounded towers. But rounded towers are also ubiquitous in Armenian and Muslim fortifications of the twelfth and eleventh centuries and before. There is therefore no particular reason to see them as derived from the West, especially as many of the masons who would have been required at Crac to cut the curved ashlars for building them would have been Easterners. Both these castles have double walls defended by projecting towers; in this respect they are similar to 'Atlīt, Ṣafad and Ṭarṭūs and more 'advanced' than Château Gaillard or any of the castles constructed by Philip Augustus. The outer wall of Carcassonne, built by 1240, represents perhaps a closer Western point of comparison. But in northern Europe this type of 'concentric' planning was not fully developed until the last quarter of the thirteenth century.[32]

Nor can we accept the view that the glacis or talus constructed against the inner wall at Crac represented a point of weakness in the design (cf. p. 79), particularly as it would have been topped by a high vertical wall to prevent escalade. Its principal function was evidently to provide additional buttressing to the wall in the event of enemy mining or a repetition of the earthquake which had destroyed much of the earlier castle in 1170; it also allowed the defenders on the wall-head a raking field of fire across its face. A particular feature of this glacis, which apparently escaped Lawrence's notice, was a *chemin de ronde* inside it, which gave access to a series of casemated arrow-slits. These probably correspond in function with the *crotae* or *fortie cooperte* recorded at Ṣafad in 1260,[33] and would have allowed archers to control the lists from

[31] C. N. Johns, *Guide to 'Atlit* (Jerusalem, 1947), 52–5, 77–81, figs. 16, 27.

[32] See P. Héliot, 'Le château de Saint-Gobain et les châteaux de plan concentrique en Europe occidentale', *Gladius*, 12 (1974), 43–58.

[33] R. B. C. Huygens, *De constructione castri Saphet: Construction et fonctions d'un château fort franc en Terre Sainte* (Amsterdam, 1981), p. 40, ll. 177–9. The same feature is found in the 13th-century glacis added to the rectangular keep at Ṭarṭūs: see M. Braune, 'Die mittelalterlichen Befestigungen der Stadt Tortosa/Ṭarṭūs', *Damaszener Mitteilungen*, 2 (1985), 45–54 (pp. 52–3, pl. 21).

positions of relative safety. The feature is unknown in northern Europe in this period, though it is found in Frederick II's castle-palace of Lucera in Apulia, built from 1235 onwards.[34]

Three varieties of machicolation may also be seen at Crac. The buttress-machicolations on tower P (Fig. 56) are probably derived, as Lawrence implies in a marginal note (p. 83), from Islamic types such as those displayed at Ukhaydir in the late eighth century. The box type seen on the outer west wall is already found on Syrian tower-houses of the sixth century, on the Muslim *ribāṭ* of Burj Yunqa in Tunisia in the ninth century and on the walls of Cairo in the eleventh;[35] its derivation therefore seems also likely to be Islamic. The continuous gallery-machicolation first appears at Crac in a later thirteenth-century Frankish addition to the top of tower P; those crowning the south wall of the castle, however, seem more likely to be early Mamluk. The idea that machicolations were a Hospitaller 'import' into Syria from the West should therefore be abandoned.[36]

Thus far there is therefore nothing in the Hospitallers' architecture at Crac or Marqab to distinguish it in any functional sense from that of the Templars, save perhaps the common use made of rounded as opposed to rectangular towers. We have already noted, however, that the Templars did on occasion build rounded towers; and, as will be shown, a number of Hospitaller castles were built with rectangular ones. One such is their castle of Belvoir, constructed between 1168 and 1187. C. H. C. Pirie-Gordon's plan, made in 1908, shows this as a rectangular enceinte, flanked by rectangular corner- and interval-towers (Fig. 47). Excavations in the 1960s, however, have revealed an inner ward comprising another rectangular enceinte with four rectangular towers at the corners and a fifth containing a bent entrance on the west.[37] Unfortunately little more than the foundations and lower parts of the walls are left, following the castle's destruction in 1218. The towers of the inner ward, however, can be seen to have contained arrow-slits providing flanking fire along the faces of the inner curtain walls. Those of the outer ward also had postern gates set in their sloping side walls, giving into the bottom of the moat to enable the defenders to sally out and set fire to the enemy's siege-towers and engines. After the battle of

[34] A. Haseloff, *Die Bauten der Hohenstaufen in Unteritalien* (Leipzig, 1920), i, figs. 17–18, 30–2, 36, pls. VIII, X–XII; C. Shearer, *The Renaissance of Architecture in Southern Italy: A Study of Frederick II of Hohenstaufen and the Capua Triumphator Archway and Towers* (Cambridge, 1935), 156–7; P. Héliot, 'Un organe peu connu de la fortification médiévale: la gaine', *Gladius*, 10 (1972), 45–67 (pp. 64–6, fig. 12).

[35] See K. A. C. Creswell, *Early Muslim Architecture*, i. Umayyads, A.D. 622–750, 2 vols. (Oxford, 1932, 2nd edn. 1969), ii. 540–2; Pringle, *Defence of Byzantine Africa*, ii, pl. LXXXIa. And above, n. 10.

[36] See Deschamps, *Châteaux*, i. 262–6.

[37] For notes, see p. 66.

Ḥaṭṭīn in 1187, the Hospitaller garrison of Belvoir endured eighteen months of siege before finally accepting a safe conduct to Tyre in January 1189.

Castles such as Hospitaller Belvoir or Templar La Fève, with rectangular plans and two wards, one inside the other, do not appear to have been uncommon in the Crusader states in the twelfth century. Bayt Jibrīn, a twelfth-century Hospitaller castle now undergoing archaeological study, appears to have been similar to Belvoir in this respect.[38] William of Tyre describes it in 1136 as 'a strong fortress surrounded by an impregnable wall with towers, ramparts, and a moat',[39] though it is uncertain whether it already had an outer wall by this date. Other rectangular castles with towers at the corners are also recorded by William of Tyre. Yibnā (Ibelin), for instance, was built in 1144 on top of a hill, 'of very strong masonry with deep foundations and four towers';[40] and Blanchegarde (Tall al-Ṣāfī), built the same year, was 'a stronghold of hewn stone, resting on solid foundations . . . adorned with four towers of suitable height'.[41]

It was an easy matter to convert a rectangular castle of the type of Yibnā or Blanchegarde into a concentric castle like Belvoir. The process may be accurately documented at Darum (Dayr al-Balaḥ) south of Gaza. Here by 1170, King Amaury had constructed 'a fortress of moderate dimensions, covering scarcely more than a stone's throw of ground. It was square in form and at each corner was a tower, one of which was more massive and better fortified than the rest. There was neither moat nor barbican (antemurali).'[42] In 1170, Saladin's forces succeeded in breaking into the castle and, after burning down the door, occupied the lower part of the larger tower, or keep, while the defending garrison retreated to the upper floor (which evidently was of stone). The castle did not fall on this occasion, however, and it was evidently soon provided with additional defences to prevent a recurrence. In 1187, however, it was surrendered to Saladin after the battle of Ḥaṭṭīn and four years later the roles were reversed when the newly installed

---

[38] Although the excavator unaccountably appears to think that the castle is Mamluk, even though the 12th-century church is plainly built up against its inner ward: cf. A. Kloner and D. Chen, 'Bet Govrin: Crusader Church and Fortifications', *Excavations and Surveys in Israel*, 2 (1983), 12–13. See also Benvenisti, *Crusaders*, pp. 185–8.

[39] *Historia*, xiv, ch. 22: *Recueil des historiens des croisades: Historiens occidentaux* (Paris, 1844), i. 639 (trans. E. A. Babcock and A. C. Krey, *A History of Deeds Done Beyond the Seas*, 2 vols. (New York, 1943), ii. 81).

[40] *Historia* xv, ch. 24: *RHC Occ.* i. 696 (trans. ii. 130).

[41] *Historia* xv, ch. 25: *RHC Occ.* i. 698 (trans. ii. 131). The inner ward of the contemporary Hospitaller castle of Belmont also seems to have had a rectangular plan, though the outer ward followed the contours of the hill on which it stood: cf. R. P. Harper and D. Pringle, 'Belmont Castle: A Historical Notice and Preliminary Report of Excavations in 1986', *Levant*, 20 (1988).

[42] William of Tyre, *Historia*, xx, ch. 19: *RHC Occ.* i. 973 (trans. ii. 372–3).

Muslim garrison found themselves besieged by King Richard I. The account of Richard's storming of the castle allows us to assess the changes that had been made between 1170 and 1191 (or 1187).

In the castle of Darum there stood seventeen very strong and appropriately positioned towers, one of which was stronger and more prominent than the others; this was surrounded by an outer and deeper ditch, which moreover was consolidated on one side with paving stones [i.e. presumably a masonry glacis], while on the other there projected the natural rock.[43]

It is a matter of simple arithmetic to deduce that to the earlier castle of four towers had now been added an outer enceinte with a tower at each corner and two or three interval-towers or turrets on each side.

As Lawrence notes, rectangular castles such as Yibnā and Blanchegarde have more in common with the fortifications of the sixth-century Byzantine frontier region than with anything being constructed in the twelfth century in the West (p. 68). More obvious prototypes, however, lay nearer at hand in the early Muslim forts of Mīnat al-Qal'a (Castellum Beroart to the Crusaders), north of Ascalon, and Kafr Lām (Cafarlet), south of Hayfā; both of these existed by the end of the tenth century.[44] Were it not for certain diagnostic features, such as the solid cylindrical turrets at their corners and flanking their gates and the rectangular buttresses projecting from the face of their curtains, these might easily be (and indeed occasionally have been)[45] mistaken for Frankish constructions. 'Concentric' planning with an outer wall and ditch is also found in the Seljuk defences of Jerusalem, which the Crusaders stormed in 1099,[46] and in the Fatimid walls of Ascalon which they took in 1153.[47] By 1212, the Franks had surrounded Beirut and Acre with double walls on their landward sides;[48] and Tyre already had

---

[43] *Itinerarium peregrinorum et gesta Regis Ricardi*, v. 39: ed. W. Stubbs (Rolls Series, 38. 1; London, 1864), 353. The French verse version of the same events by Ambroise differs only in specifying that the 17 towers included 'tors et toureles': cf. G. Paris (ed.), *L'Estoire de la guerre sainte (1190–1192)* (Paris, 1879), ll. 9223–9.

[44] Benvenisti, *Crusaders*, pp. 326–31; A. Kloner, 'History of Ashdod Yam', *Nature and Land*, 16 (1973–4), 21–4 [Hebrew]; A. El'ad, 'The Coastal Cities of Palestine During the Early Middle Ages', *Jerusalem Cathedra*, 2(1982), 146–67. Mīnat al-Qal'a, or Māhūz Azdūd, is mentioned by al-Muqaddasī (c.985) as one of the coastal stations where Christian prisoners were ransomed by the Byzantines: cf. trans. by G. Le Strange, *Palestine Pilgrims Texts Society Library*, 3 (London, 1892), 62.

[45] Thus, Smail, *Crusading Warfare*, pp. 231, 235; S. Langé, *Architettura delle crociate in Palestina* (Como, 1965), 32, 108, 120, 181, fig. 43.

[46] See J. Prawer, 'The Jerusalem the Crusaders Captured: A Contribution to the Medieval Topography of the City', in P. W. Edbury (ed.), *Crusade and Settlement* (Cardiff, 1985), 1–16 (pp. 1–5). The building of the walls began around 1033, and continued through the century.

[47] D. Pringle, 'King Richard I and the Walls of Ascalon', *Palestine Exploration Quarterly*, 116 (1984), 133–47 (pp. 134–6).

[48] Wilbrand of Oldenburg, *Itinerarium Terrae Sanctae*, i. 1, i. 5: ed. J. C. M. Laurent, *Peregrinatores Medii Aeui Quatuor* (Leipzig, 1864), 163, 166; cf. D. Jacoby, 'Montmusard,

double walls to seaward and triple walls to landward when it fell to the Crusaders in 1124.[49]

Turning to consider developments in castle-building in France and England in the twelfth century,[50] it may now be possible to detect similarities and contrasts with the types of castle being built by the Franks of Outremer in the same period. As far as the development of keeps was concerned, it is possible to agree with Lawrence that there seem to be no grounds for believing that the changes in design and the final adoption in France, and later in England, of the cylindrical keep as the preferred form, were anything other than purely indigenous European developments. The only features of keeps for which oriental precedence suggests the East as a likely source of inspiration are the use made of machicolations and of a masonry talus to protect the base of the wall.[51]

Curtain walls containing a double-tiered *chemin de ronde* are found in the Byzantine town wall of Ruṣāfa in the sixth century, and their appearance at Crusader Ṣahyūn in the twelfth century may also have been influenced by surviving Byzantine examples, perhaps at the same site.[52] There is no need to explain the examples in the West as the result of Crusader influence, however, for examples are noted at Mont St Michel and the keep of Loches castle by the end of the eleventh century, and a possible model already existed in sections of the walls of Rome, constructed under the Emperors Aurelian and Maxentius in the late third and early fourth centuries AD.[53] Arrow-slits with sloping sills appear at Crac des Chevaliers and in England in the 1170s, and in France from around 1200;[54] there is therefore insufficient evidence to suggest an Eastern origin for the Western types. More research is

Suburb of Crusader Acre: The First Stage of its Development', in B. Z. Kedar, H. E. Mayer and R. C. Smail (eds.), *Outremer* (Jerusalem, 1982), 205–17 (pp. 211–17).

[49] William of Tyre, *Historia*, xiii, ch. 5: *RHC Occ.* i. 562 (trans. ii. 9); cf. Burchard of Mount Sion, *Descriptio Terrae Sanctae*, ii. 5: ed. Laurent, *Peregrinatores*, p. 25 (trans. A. Stewart, *Palestine Pilgrims Texts Society Library* (London, 1896), 9, 11–12).

[50] For a general survey, see P. Héliot, 'Le Château-Gaillard et les forteresses des XIIᵉ et XIIIᵉ siècles en Europe occidentale', *Château Gaillard: Études de castellologie européenne*, 1 (Caen, 1964), 53–75.

[51] P. Héliot, 'La genèse des châteaux de plan quadrangulaire en France et en Angleterre', *Bulletin de la Société nationale des antiquaires de France* (1965), 238–57 (p. 254); id., 'L'évolution du donjon dans le nord-ouest de la France et en Angleterre au XIIᵉ siècle', *Bulletin archéologique du Comité des Travaux historiques*, NS 5 (1969), 141–94 (pp. 184, 190); id., *Château Gaillard*, (1964), 1 64–5, 67; Finò, *Forteresses*, pp. 163–7; A. Mersier, 'Hourds et machicoulis', *Bulletin monumental*, 82 (1923), 117–29.

[52] Smail, *Crusading Warfare*, p. 241.

[53] Héliot, *Gladius*, 10 (1972), 60–7; cf. I. A. Richmond, *The City Wall of Imperial Rome* (Oxford, 1930), 65–72, figs. 3, 8–10; M. Todd, *The Walls of Rome* (London, 1978), 30–1, 49, figs. 10, 11, 22–3.

[54] Héliot, *Bull. archéol. du Comité*, NS 5 (1969), 182–4; Finò, *Forteresses*, pp. 163; 218–20.

needed on the types of gateways built by the Crusaders in the twelfth century before one can tell whether the systems that they incorporated were more or less advanced than contemporary ones in the West. By the end of the century, use was being made in East and West of portcullises, wing-doors and machicolations in varying combinations, and the systems were to become more elaborate by the end of the following century.[55] One particular oriental feature of which the Crusaders made use in the twelfth and thirteenth centuries but which seems never to have been much favoured in the West, was the bent entrance set in the side wall of a projecting tower.[56]

The two most significant 'progressive' features of Crusader castle-building in the twelfth century that were to become increasingly common in the West only in the thirteenth century, however, were 'rectangular' and 'concentric' planning. As we have seen, rectangular castles with towers placed at their corners in the manner of late Roman forts were being built by the Franks of Outremer in the 1130s, probably copying local Muslim prototypes. In France, castles of this kind, but more often with cylindrical towers, were not built until the last quarter of the century. The series begins with Druyes-les-Belles-Fontaines in Auxerrois (1170/1200) and Châtel-Guyon (Basse Auvergne), and continues in the castles of Philip Augustus: Yèvre-le-Châtel (c.1200), the Louvre in Paris (c.1200), Caen (1204/5), Péronne in Vermandois (1205/10), St Omer (Artois) (c.1212) and Montreuil-sur-Mer (Ponthieu) (−1224). Such castles were unusual in Britain before the end of the thirteenth century.[57] Various sources of inspiration have been suggested, including the example of surviving Roman forts, the reading of Roman military and architectural treatises and the influence of the Crusades. Neither Vegetius nor Vitruvius advises the use of a rectangular plan, however, and among Gallo-Roman town defences irregular or oval plans by far outnumber rectilinear layouts.[58]

Concentric planning such as is found at Darum and Belvoir between 1170 and 1187 and in the walls of Tyre, captured in 1124, and of Ascalon, taken in 1153, is not found in France before the second

[55] See P. Deschamps, 'Les entrées des châteaux des croisés en Syrie et leurs défenses', Syria, 13 (1932), 369–87, pls. LXXVII–LXXXVI; J. Mesqui, 'La fortification des portes avant la Guerre de Cent Ans: essai de typologie des défenses des ouvrages d'entrée avant 1350', Archéologie médiévale, 11 (1981), 203–29.

[56] Deschamps, Syria, 13 (1932), 372–7; Pringle, Defence of Byzantine Africa, i. 170. It has been suggested that the bent entrance in the Horseshoe Gate at Pembroke Castle was the direct result of oriental influence. Its builder William Marshal, Earl of Pembroke, had spent 2 years in the Holy Land after 1183, attached to the Templars: see D. J. Cathcart-King, 'Pembroke Castle: Derivations and Relationships of the Domed Vault of the Donjon and of the Horseshoe Gate', Château Gaillard: Etudes de castellologie médiévale, 8 (Caen, 1977), 159–69.

[57] Héliot, Bull. de la Soc. nat. des antiquaires de France (1965), 239–51.

[58] Ibid. 251–7.

quarter of the thirteenth century. The outer wall at Carcassonne may be dated to 1228–39, though it was repaired later in the same century;[59] and Saint Gobain in the Laonnais seems to have been a rectangular castle with five towers and an outer apron-wall containing a *chemin de ronde* built before 1250.[60] The system is not fully developed, however, until the construction of King Edward I's Welsh castles of Rhuddlan (1277–82), Harlech (1283–90) and Beaumaris (1294/5–), and those of his vassals at Kidwelly (*c.*1275–) and Caerphilly (1271–).[61]

To conclude, it seems unlikely that any definitive answer to the question of East–West influences in medieval castle-building will be possible until Crusader, Armenian, Muslim, Byzantine and, one should add, Italian and perhaps Spanish castles and town defences have been subjected to the same kind of scrutiny that English, Scottish, Welsh and French ones have undergone in the last seventy-five years. And even then it may well turn out that arguments for external stimulus and for indigenous development are not mutually exclusive. Indeed, international rivalry in military technology in the modern world illustrates how remarkably similar weapon systems, and counter-systems, can be developed by opposing sides apparently quite independently.

Such as it is, the evidence reviewed here leads us to agree with Lawrence that the planning of many early twelfth-century Crusader castles owed more to the West than to the East and that twelfth-century castles built in the West, or at least in France and England, owed more to the native genius of their builders than to external influences from the East. But, this said, it also has to be recognised that, by the end of the twelfth century, techniques of fortification being practised in the Crusader states of Outremer were far in advance of those being used in the West; and it was not until the end of the thirteenth century that the West had anything to compare with Belvoir, ʿAtlīt, Ṣafad or Ṭarṭūs.

A much stronger case can be made, however, for oriental techniques having influenced castle-builders in the West in the thirteenth century. In addition to the Crusader fortifications which still survive, it is also important to consider some of the more significant ones that do not. Whether or not visiting Westerners ever saw Crac, Marqab, or Ṣafad during their stay in the Holy Land, most would certainly have inspected the walls of Tyre and Acre. One such visitor in 1271–2 was Prince Edward of England, later to become King Edward I. During his time in

[59] See p. 11, n. 14.

[60] Héliot, *Gladius*, 12 (1974), 43–6, fig. 1.

[61] Brown, *English Castles*, pp. 100–15; A. J. Taylor, *The King's Works in Wales* (HMSO; London, 1974); id., *Rhuddlan Castle* (HMSO; London, 1956; repr. 1972); id., *Harlech Castle* (Cadw; Cardiff, 1985); id., *Beaumaris Castle* (HMSO; Cardiff, 1980); J. R. Kenyon, *Kidwelly Castle* (Cadw; Cardiff, 1986); C. N. Johns, *Caerphilly Castle* (HMSO; Cardiff, 1978).

Acre, it seems that the Prince not only saw but actually contributed to improving the defences of the city. For when Acre fell to the Mamluks twenty years later, the 'Barbican of the Lord Edward' (*sbaralium domini Odoardi*) was one of the formidable stone-built outworks which the Muslims had to undermine and capture before they could begin to approach the outer walls of the city.[62]

[62] Marino Sanudo, *Liber Secretorum Fidelium Crucis*, ed. J. Bongars (Hanau, 1611, repr. Jerusalem, 1972), 230; cf. J. Prawer, *Histoire du royaume latin de Jérusalem*, 2 vols. (2nd edn. Paris, 1975), ii. 555 n. 41; G. Schlumberger, *Prise de Saint-Jean-d'Acre en l'an 1291* (Paris, 1914), 39; Deschamps, *Châteaux*, i. 69; id., *Syria*, 13 (1932), 387. The half-moon barbican of the Tower of London was also the work of Edward I (cf. Brown, *English Castles*, pp. 116–18). That certain features of Caernarvon Castle, such as the banding of the masonry and the use of polygonal towers, were conscious imitations of the Theodosian walls of Constantinople is now well established, though in this instance the imitation had a symbolic rather than a defensive purpose: see Taylor, in *History of the King's Works*, i. 370–1; Brown, *English Castles*, p. 102.

CRUSADER CASTLES

THE INFLUENCE OF THE CRUSADES ON EUROPEAN MILITARY ARCHITECTURE TO THE END OF THE TWELFTH CENTURY

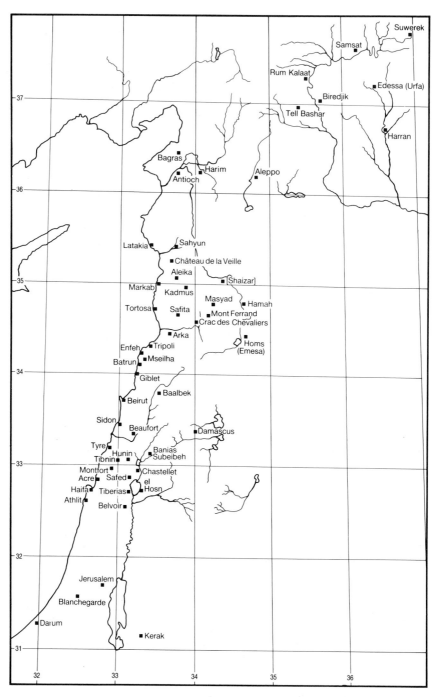

Map I. Syria and Edessa

I. APOLOGY FOR THE WORK
AND OBJECTION TO SECONDARY CRITICISM

ANY consideration of the influences of Levantine military architecture on that of the West must almost of necessity be minute and technical: and any such inquiry must obviously be based on first-hand study of the actual remains of twelfth-century castles, in Syria and Europe.[1] A few of the castles in the East have been adequately described with plans and illustrations, but beyond these there are many, often of equal importance, of which details have never been published; and the sites of some, which figure in history, remain unidentified in the riot of hills filling up Syria between Antioch and Nazareth. Reference is made here to some forty Crusading castles, including, for the twelfth century, nearly all those in the East. The materials for the thirteenth century in Armenia and the Greek islands are almost entirely unworked; there has, as a matter of fact, been practically no exhaustive study even of the castles of that period in Europe.

II. A PRELIMINARY
TO SET FORTH OMAN'S VIEW, AND TO SUGGEST THE LINES ON WHICH I WILL WORK

THE classical view of the subject may be summed up in the statement that 'the Western builders . . . were for many years timid copyists of the

Plans have been taken from published sources, which are mentioned. I am indebted to Mr Pirie-Gordon for the plans of Banias (an exceptionally fine plan), Hunin, Belvoir (Kaukab el Howa), Harim, Antioch, Areimeh, and Kafr-lam, all of which were made by himself in Syria in 1908, and are at present unpublished. The remaining plans are my own work. The spelling of Arabic names usually is French or German, according to the source from which they are taken. Some are my own.

In quoting places for the first time the name is spelt first in the common European form (if one exists). If not, the Frank name from some original authority is given. After the first appearance the place may be called by any of its names.

Violently controversial points are usually settled by a plain assertion, for simplicity and peace. If they are of importance in my argument they may be discussed.

[1] The Byzantine castles in northern Africa are described from Diehl. Except for this, no castle has been mentioned unless personally visited. This will account for the omission of a few well-known 12th-century castles, such as Keràk or Grignols.

crusading architects'.[2] The idea is that the Franks marched East with hardly any understanding of fortifications more elaborate than earthworks; and that in their passage through Roumelia and Asia Minor they were so dazzled with the architectural genius of the Greeks that they laid aside their rudimentary ideas of defensive work in favour of a wholesale parody of the castles of the Eastern empire, 'learning everything and forgetting nothing'. Their supreme contempt for the light-armed Greeks, who could not, or at least did not, wear the heavy armour of the Frankish knights—until Manuel Comnenus beat them on their own ground—enabled them to appreciate the assistance lent the weaklings by fortresses in their prolonged resistance against attack on three sides. And having thus turned their attention to the military architecture of the Byzantines, they soon discovered its peculiar suitability to the conditions of Eastern warfare.[3] The Crusaders, therefore, copied the Greeks, whom they despised; and the precarious situation of the Latin kingdom made it the more necessary that means should be found, as in the Eastern Empire, to restore the balance between defence and offence. In Europe in the same way it is said that all the excellences of Western castle-building are due to the quickening instance of the later Greek Empire.[4] Professor Oman quotes 'outer wards and foreworks' . . . with their 'numerous and strong curtain towers', as borrowings from Byzantium.[5] Another imported feature is the provision of flanking or covering fire for exposed points, with the general idea of 'concentric' castles. Even Château Gaillard, the masterpiece of Richard I, is supposed to have drawn the greater part of its excellences from the East; sometimes it has been said[6] that Syrian workmen were imported to build it; at least that Richard was incapable of it before his experience in Palestine.

On the other hand, examination of Crusading castles in Syria itself, and a comparison of them with contemporary castles in France appear to lead to conclusions wholly different. It is obvious that in the early state of the Latin kingdom, the period of the private feudatories, the castles erected in Syria were of a purely Western pattern. Later on, the two great Orders developed rival styles, of which one, that of the Hospital, drew its inspiration from Europe, and the other, that of the Temple, from the Byzantine Empire. No castles of the Templar type were erected in France before the general adoption of the use of gunpowder.[7] Castles of Byzantine characteristics, cited in standard

---

[2] The text is taken from Oman's [History of the] Art of War[, p. 532], which apparently is the recognised authority of Oxford dons on medieval architecture: and is altogether futile. [R]

[3] i.e. absence of woodwork; but all this is supposed to explain Oman's mentality. [X]

[4] All this is Oman. [RX]   [5] [History of the] Art of War, p. 533.

[6] This is Mrs [Eugénie] Strong. [RX] [Reference obscure.]

[7] And very few after! [RX]

works, date from the fifteenth and seventeenth centuries, when the Crusades were well-nigh forgotten, and the Syrian fortresses for two centuries and more in the hands of the Infidel. Unless nearly contemporary instances of borrowing of Byzantine principles can be found, the classical view can hardly be accepted; a mere transfer of some trifling detail need cause no surprise, for there was constant interchange between East and West. There was no important family in southern France or northern Italy which had not a younger branch in the Levant, and these younger branches died out so quickly that there was a continual traffic of the higher classes, concerned about material possessions, quite apart from the pilgrim-fervour of the rank and file.

Nor will it be enough to find curtain-towers and provision for flanking fire, and outer wards, in French castles, and cry them up as Eastern features, without troubling to search for them as integral parts of Greek fortresses; and to decide that Château Gaillard must have been inspired by the East, simply because it is superior to the general style of European fortresses, without quoting parallels in Syria, is hardly convincing.

Obviously, in Europe, France is the country chiefly to be considered. In the Middle Ages, she produced all that was best in Gothic art. Italy flowered later, after the downfall of medieval culture;[8] and in the twelfth and thirteenth centuries, Germany also was barbarous in art, and had chivalry by no means equal to that of the Western kingdoms. From France and England come all medieval masterpieces in literature and architecture; Italy had a hybrid civilisation, much more tinged than that of France with Greek and Saracenic influences, and apparently her style of fortification shows distinct Byzantine feeling.[9]

By putting side by side the development in fortification of Western Europe and that of the Eastern Empire to the beginning of the twelfth century, it should be possible to distinguish the debt owed respectively to each by Syrian architects, and a comparison of the essentials in style of a large number of Syrian castles will show their contrast with the principles in fashion in Europe at the same period. To consider the question with a knowledge of only one or two is not sufficient.[10] The buildings of the three countries are equally distinct and equally important; and many of the vague present-day theories are generalisations on insufficient material, simply through non-recognition of this fact.

[8] Apologies for not having been to Italy: an attempt to show it doesn't matter. [R]
[9] I haven't been there yet. [X]
[10] This is partly for Rey: he does not know French architecture. [X]

III. MILITARY ARCHITECTURE

IN EUROPE BEFORE THE FIRST CRUSADE

IN Britain the invasion of the Saxons meant the burning and laying waste of the walled cities of the half-Romanised inhabitants. The Saxons had a horror of living within stone walls; and examples such as the sack of Anderida [Pevensey] quite well account for their peopling their sites in imagination with devils.[1] In Gaul, on the other hand, the collapse of the Roman Empire was before barbarians, who had for generations served in her armies, and whose great ambition was to adopt her customs, and manners, and titles. Therefore they preserved her public buildings, and the Gallo-Roman population lived as before within their towns.

The more important Roman stations in Gaul had been carefully stone-walled, usually on a rectangular plan, the translation in stone of the earth-mounds of their entrenched camps. The mason work was of grouted rubble, ashlar-faced, built in sections,[2] with at intervals (generally rather long intervals) half-round curtain-towers set upon them. Sometimes a double wall was built, and the interval packed with earth, to the level of the top of the inner wall. The outer one would then form a parapet as at Toulouse.[3] The curtain-towers were generally quite small in diameter, and projected only a true semi-circle beyond the wall. At times their bases would be rectangular or polygonal (Fig. 4); only very rarely were they rectangular themselves. The gates were flanked by semi-round towers upon occasion, as at Richborough, but they are rarely so symmetrical.[4] In Britain the Colchester type of gate is

[1] [Pevensey fell to the Saxons in 491, no Briton being left alive: cf. *The Anglo-Saxon Chronicle*, trans. D. Whitelock (London, 1961), 11. The idea of the Saxons' avoidance of Roman sites still persists (cf. D. M. Wilson (ed.), *The Archaeology of Anglo-Saxon England* (London, 1976), 7–8), despite a growing body of archaeological evidence showing continued occupation of former Romano-British towns (see M. Biddle, ibid. 103–12).]

[2] This is why Roman walls crack every 40 yards, e.g. Pevensey. [X]

[3] [Viollet-le-Duc, pp. 9–10, fig. 5 = *Dictionnaire raisonné*, i. 331. The wall is now tentatively dated to the late 2nd or early 3rd century: cf. P.-A. Février, 'The Origin and Growth of the Cities of Southern Gaul to the Third Century A.D.', *Journal of Roman Studies*, 63 (1973), 1–28 (p. 27); M. Labrousse, *Toulouse antique* (Bibl. des Écoles françaises d'Athènes et de Rome, 212; Paris, 1968), 237–90.]

[4] [The main gate at Richborough (c.275–85) was flanked by rectangular towers (S. Johnson, *The Roman Forts of the Saxon Shore* (London, 1976), 49, 121, figs. 30, 66). Lawrence may have been thinking of Pevensey, where the late 3rd- or 4th-century west gate is indeed of the type described (Johnson, pp. 58, 121, figs. 36, 66, 68).]

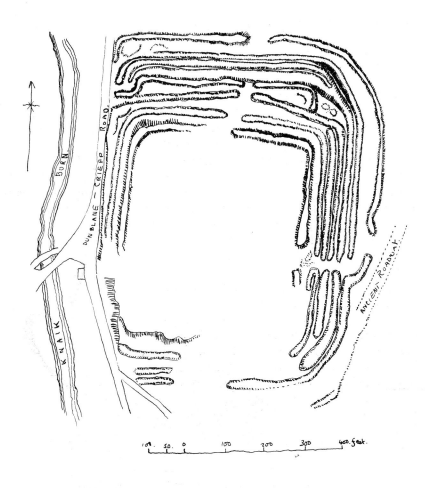

1. Ardoch, Perthshire (Allcroft, p. 331). A Roman station with six banks

a common one:[5] abroad often they were hardly defended at all.[6] Before the walls was usually a ditch of some depth and width: in exposed camps in Britain are found half a dozen ditches, with breastworks, probably palisaded, lining them.[7]

To judge from the account of the thirteenth-century siege of Carcassonne there had there been more than one line of concentric Roman wall,[8] but as a rule the one lofty stone wall and ditch were sufficient ward against the uncivilised enemies, without siege-trains or settled discipline, with which Rome had to deal: the greater part of the forts were only sure bases on a frontier, from which her troops could carry out the vigorous offensive which was her defence. The majority of the town walls in the interior of France date from the times of the later empire, and that they were put up in haste is shown by the choice of material in their construction. The walls of Narbonne, Tours, Auxerre, Orléans, have been worked in modern times as a quarry of fragments of frieze, cippi, and capitals of columns.[9]

The account of Gregory of Tours leaves no doubt that they were maintained by their later inhabitants in a fit state of defence. Describing the city of Dijon he says that it was

a fortress with very strong walls, lying in the midst of a very fertile plain . . . On the south lies the river Oscara (Ouche), on the north another small stream, which enters in at one gate, and, passing under a bridge, goes out through another gate, tracing its sluggish course round the whole circuit of the walls . . . There are four gates towards the four corners of the world and 33 towers adorn

[5] [The allusion is to the mid-2nd-century Balkerne Gate of Colchester, which Lawrence visited with his father in Aug. 1905 (*Home Letters*, pp. 3–4, fig.). The plan of the gate was clarified by the excavations of Major R. E. M. (later Sir Mortimer) Wheeler in 1917. It consisted of a pair of arched carriageways, over 17 ft. wide, flanked by two narrower footways, set in a broad projecting gate-house with rounded sides. See M. R. Hull, *Roman Colchester* (Reports of the Research Committee of the Society of Antiquaries of London, 20; Oxford, 1958), 16–21, fig. 4, pl. II.]

[6] e.g. Damascus. [R] May have been outworks. [X] [The Roman gates in Damascus, Bāb al-Sharqī and Bāb al-Sghīr, date apparently from the late 2nd or early 3rd century: see G. Watzinger and K. Wulzinger, *Damaskus: Die Antike Stadt* (Berlin-Leipzig, 1921), 65–77, figs. 38–45; J. Sauvaget, *Les Monuments historiques de Damas* (Beirut, 1932), 4; A. Rihawi, *Damascus: Its History, Development and Artistic Heritage* (Damascus, 1977), 129–32.]

[7] As at Ardoch. [RX] [The plan shown in Fig. 1 represents the fort as it would have appeared in the Antonine period, c.143 AD: see R. G. Collingwood and I. Richmond, *The Archaeology of Roman Britain* (London, 1969), 44, fig. 14b; and D. J. Breeze, 'The Roman Forts at Ardoch', in A. Connor and D. V. Clarke (eds.), *From the Stone Age to the 'Forty-Five* (Edinburgh, 1983), 224–36.]

[8] [This now appears unlikely: see below.]

[9] [On these and other late Roman town walls in France, see R. M. Butler, 'Late Roman Town Walls in Gaul', *Archaeological Journal*, 116 (1959), 25–50; H. von Petrikovits, 'Fortifications in the North-Western Roman Empire from the Third to the Fifth Centuries A.D.', *Journal of Roman Studies*, 61 (1971), 178–218; S. Johnson, 'A Group of Late Roman City Walls in Gallia Belgica', *Britannia*, 4 (1973), 210–23.]

2. Carcassonne. Five Visigothic towers (restored) [from north]

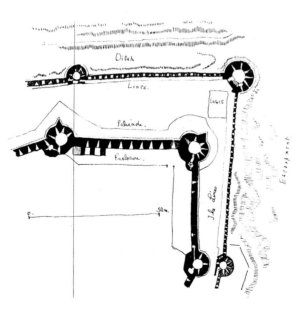

2a. An angle of the fortifications of Carcassonne [Viollet-le-Duc, *Dictionnaire raisonné*, i 379.

2b. [Carcassonne. So-called 'Visigothic' (Gallo-
Roman) and medieval wall. August 1908]

the walls. It is built of squared stones to the height of 20 feet, and above that with rubble. The height is 30 feet, and the thickness 15.[10]

Gregory often mentions fortresses, and there are still standing parts of the very remarkable walls of Carcassonne where the Visigoths or their immediate successors rebuilt the Roman enceinte after the same plan though of inferior material, on the old foundations [Fig. 2].[11] Their work was done with such thoroughness that the city proved impregnable in the early part of the thirteenth century against the determined attacks of the Trencavel party—attacks supported with all the skill in siegecraft known at the time, and in addition the most complicated scheme of mining on record. Even St Louis and Philip the Fair, when remaking the place, had nothing to alter in these defences, except the necessary repairs.[12] In the story of the siege mention is made of the 'old Saracenic (i.e. Roman) wall in the lists'.[13] This probably stood on the site of the present outer line of Philip the Fair, just as the twelfth-century castle itself, within all the walls, occupies what was undoubtedly the site of a Roman citadel, of the 'military quarter' of the town (Fig. 5).[14]

If the Visigothic city of Carcassonne cannot be called concentric, it is only because there never were any concentric castles, except here and there by accident. The aim in the mind of every architect of the smallest intelligence was to find such a site for his building that the waste and weakness of equal accessibility on all sides might be avoided: then he could multiply defences on the one weak face alone. And this he did in all ages, from the earliest earthworks to the latest fort before the introduction of siege cannon.

[10] *History of the Franks*, iii. 19 [cf. trans. L. Thorpe (Harmondsworth, 1977), 182–3. The description is confirmed by the archaeological remains: see Butler, *Arch. J.*, 116 (1959), 37.]

[11] One Roman tower (Tour de Davejean) is still standing near the Narbonne gate. (This is put in to show how Roman traditions were preserved in Provence. The 11th-century people there did not start *ab initio*. [R]) [The supposed Visigothic element in the walls of Carcassonne, however, has for long been overestimated (cf. Butler, *Arch. J.*, 116 (1959), 28–30, figs. 3–4). A major rebuilding occurred in 1228–39 (see n. 14).]

[12] The honour of improvement was reserved for Napoleon III and his architect Viollet-le-Duc who rebuilt some of them on what they imagined to be the original plan.

[13] This shows [that there was] a double line of Roman or Visigothic walls. [X] Of course Saracenic means Roman. [R] [The text in question states that the attackers in 1240 mined 'beneath a certain Saracen wall, up to the wall of the lists' (Viollet-le-Duc, p. 40 = *Dictionnaire raisonné*, i. 347). Whatever its original date or function, this 'Saracen' wall would therefore have been *outside* the outer wall or lists.]

[14] [The outer walls, which Lawrence considered to have been built by Philip the Fair (1284–1314), are now assigned to the period 1228–39 (Y. Bruand, 'La Cité de Carcassonne: Les enceintes fortifiées', *Congrès archéol. de France*, 131 (1973), 496–515); and the castle or palace, though built against the Roman wall, appears to be no earlier than the last quarter of the 12th century if it is not entirely a work of the 13th (P. Héliot, 'L'âge du château de Carcassonne', *Annales du Midi*, 78 (1966), 7–21; Y. Bruand, 'La Cité de Carcassonne: la citadelle ou château comtal', *Congrès archéol. de France*, 131 (1973), 516–31).]

Plan of the Tower
A: Ground plan
B: First-floor
C–D: Pits under drawbridge

Interior of Visigothic
tower showing the
isolated doors

3. [Carcassonne] (Viollet-le-Duc, p. 11)

4. Exterior of tower in Visigothic Carcassonne, as
restored by Viollet-le-Duc (p. 12). This tower, quite
unusually, has two stories

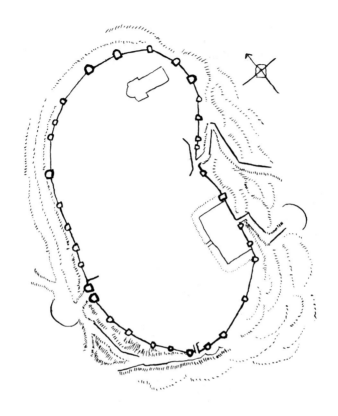

5. Visigothic Carcassonne (Viollet-le-Duc, p. 17)

The Roman, and of course the Visigothic, towers of Carcassonne are solid to a height of nearly 40 feet (cf. Fig. 4): the idea of loopholes, whose narrowness makes hollow towers nearly as secure as solid, does not seem to have occurred to classical architects. The openings at the top (and in the room under it in some towers) are large and square, to be closed with heavy hinged louvre-boards or shutters against arrows or stones from outside. These shutters were used, till *machicoulis* were invented, on the merlons of the parapet.[15] Some of the towers stand on square bases (Fig. 4) of the height of some 8 feet: and each tower was cut off from the *chemin de ronde* of its curtain by a gap which could be filled in peacetime with a movable bridge (Fig. 3). The little door of entry was also strongly barred, and so the towers are, if necessary, a series of independent fortresses. If their small size made them worthless as a last resort, at least they were effectual barriers against surprise: and they gave moral support to their defenders, who were freed from the necessity of guarding their flank and rear. It is calculated that there must have been rather more than thirty such towers in the wall of Carcassonne.[16]

A tradition of castle-building in stone existed in other parts of Europe also.[17] In Germany, Nicetius of Trier built a castle with thirty towers for the protection and oppression of his people,[18] and in Italy there were many stone castles new-built, and more adapted from Roman edifices of some sort.[19] Yet probably these may be considered grand exceptions. The ordinary fortress of the Dark Ages, indeed down to the middle of the eleventh century, was a mound, or hill-ditch cut in the soil, with nothing heavier than a palisade of tree trunks on the top. Incursions, such as those of the Northmen, were, in the ninth century, the most compelling cause of castle-building and such hurried raids were better met by earthworks than by the most scientifically planned structure in masonry. Even unskilled workmen could throw up a defensible fortress in a few days,[20] and an earthwork is by no means a thing to be lightly

---

[15] Cf. Fig. 38. Safita, 12th century. [X]

[16] [It should be noted that the towers are described here (after Viollet-le-Duc, pp. 10–13, figs. 6–8 = *Dictionnaire raisonné*, i. 332–4) as they would have appeared after the refurbishment of 1228–39.]

[17] Not only in Provence were there Roman traditions. Alfred in London used Roman walls. [R] [See M. Biddle and D. M. Hudson, *The Future of London's Past* (Worcester, 1973), 22–3.]

[18] [Bishop of Trier, 525/6–561/85. The castle and palace are described by Venantius Fortunatus, *Carminum*, iii. 12: ed. F. Leo, *Monumenta Germaniae Historica, Auctores Antiquissimi*, iv. 1 (Berlin, 1881), 64–5 (Fr. trans. in G. Fournier, *Le Château dans la France médiévale* (Paris, 1978), 263–4).]

[19] [An example, though mostly of brick rather than stone, is the wall encircling the Vatican built by Pope Leo IV in 846. See S. Gibson and B. Ward-Perkins, 'The Surviving Remains of the Leonine Wall', *Papers of the British School at Rome*, 47 (1979), 30–57; 51 (1983), 222–39.]

[20] The Nervii built lines 15 miles long, with a 9 ft. vallum [bank] and 15 ft. ditch, in 3 hours

contemned. Caesar had to proceed against one in Britain in most formal fashion, by testudo and agger,[21] and naturally, against assailants of the same quality as the defenders, such fortresses as Maiden Castle (Fig. 6) or Hembury fort [Fig. 10] were impregnable.[22] Of those of as late as the time of the Danish invasion of England, it can be said 'hardly one fell in 20 years of war',[23] and there is no reason for thinking that Alfred's *burhs* were of exceptional efficiency.

Unfortunately the modern confusion in the study of earthworks makes them almost hopeless subjects to date or even to argue about. Certain facts are known historically: such as the construction of *burgs* by Charles the Great in great numbers; the fortified bridges and camps of Charles the Bald in northern France; the *burhs* of Alfred and his family, and the earthworks of Henry the Fowler in Germany. One may infer others: that the century-long struggle of Britons and Saxons in the South Midlands could not have lasted so many years without fortresses on both sides: that if Offa made his colossal dyke merely 'to mark the limit' of his kingdom, it shows a remarkable degree of proficiency in earth-working: and that presumably these wars, and those against the Danes would leave more trace in the country than prehistoric struggles. Yet none the less, hardly a single one of these defence-works has been identified.[24] The modern archaeologist has a weird fondness for identifying every 'mount-and-bailey' fortress with Norman post-Conquest work. The Normans were near of kin to the Danes, and settled in north France. If they brought their system of fortification from Scandinavia one would imagine it to be very like the Danish. If it was copied from that of north France then presumably it was also copied in England long before 1066. In either case it is far too sweeping to ascribe every

according to Caesar (*Gallic War*, v. 42 [trans. H. J. Edwards (Loeb Classical Library; London and Cambridge, Mass., 1917), 289]).

[21] *Gallic War*, v. 9 [trans. Edwards, p. 247].

[22] [Excavation in the 1930s, however, provided evidence for identifying Maiden Castle as one of the 20 British *oppida* stormed by Vespasian and the Second Legion in AD 44: see R. E. M. Wheeler, *Excavations at Maiden Castle* (Reports of the Research Committee of the Society of Antiquaries of London, 12; Oxford, 1943). On the Iron Age fort of Hembury, see A. Fox, 'Hembury Hill-fort', *Archaeological Journal*, 114 (1957), 114–47; A. H. A. Hogg, *Hill-Forts of Britain* (London, 1975), 222–4; M. Todd, 'Excavations at Hembury (Devon), 1980–83: A Summary Report', *Antiquaries Journal*, 64. 2 (1984), 257–68.]

[23] Oman[, p. 111.] [R]

[24] [The situation is somewhat different today. See, for instance, N. Brooks, 'The Unidentified Forts of the Burghal Hidage', *Medieval Archaeology*, 8 (1964), 74–90; C. A. R. Radford, 'The Later Pre-Conquest Boroughs and their Defences', *Medieval Archaeology*, 14 (1970), 83–103; J. M. Hassall and D. Hill, 'Pont de l'Arche: Frankish Influence on the West Saxon Burh?', *Archaeological Journal*, 127 (1970), 188–95; F. Kitchen, '*The Burghal Hidage*: Towards the Identification of *Eorpeburnan*', *Medieval Archaeology*, 28 (1984), 175–8; C. Fox, *Offa's Dyke* (London, 1955); and the various entries in 'Medieval Britain in ——', published annually in *Medieval Archaeology*, 1– (1957–).]

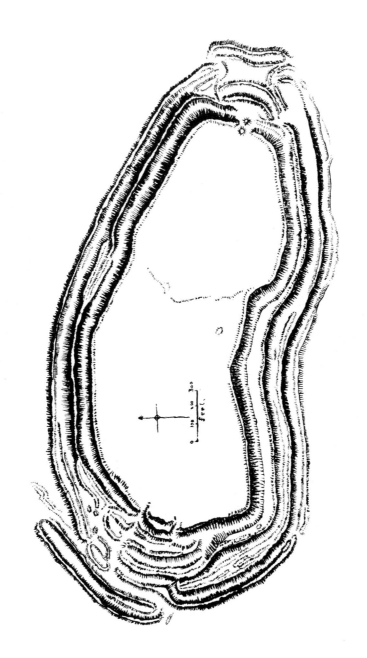

6. Maiden Castle, Dorsetshire (Allcroft, p. 101)

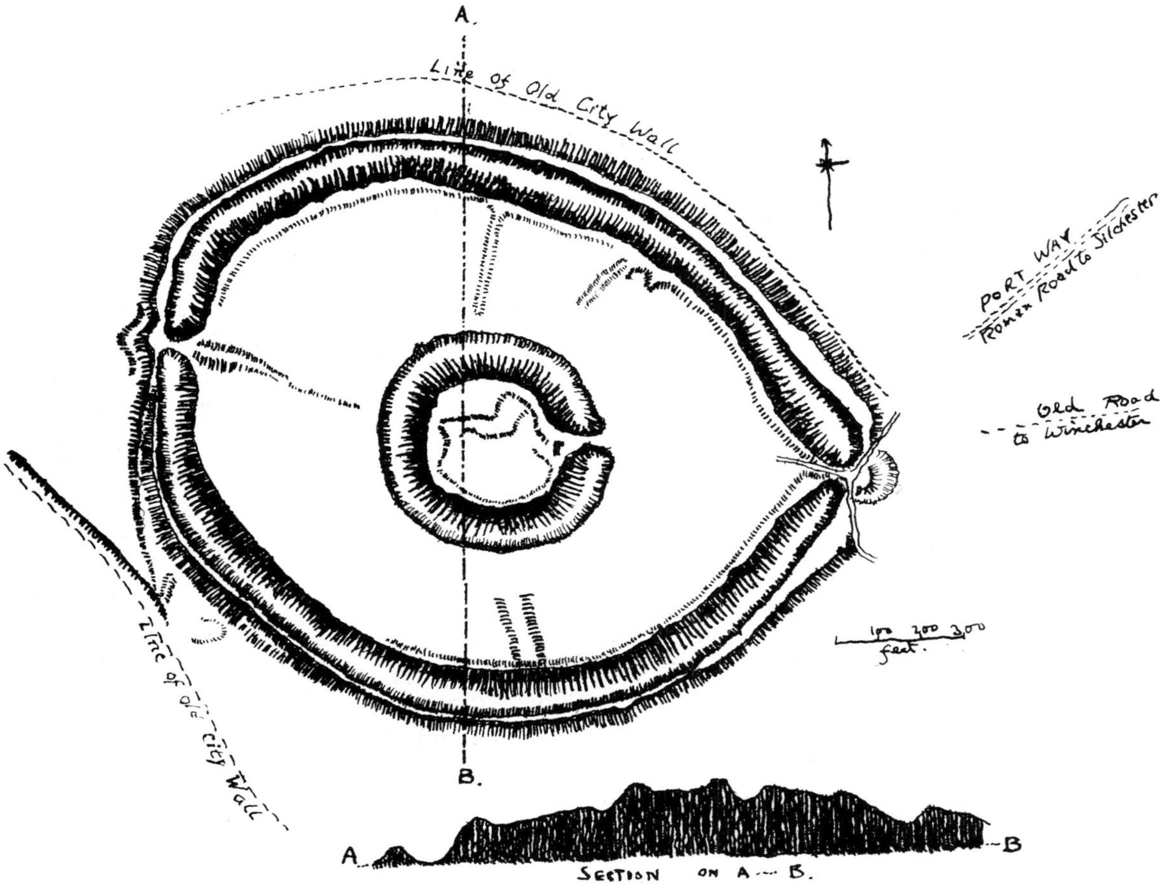

Line of Old City Wall

ROMAN ROAD TO SORVIODUNUM

Old Road to Winchester

1" 100 200 300 feet

A.

B.

Line of old city Wall

SECTION ON A — B.

A.

B.

7. Old Sarum (Allcroft, p. 119)

mound-and-bailey work to Norman hands, especially as it gives them the absurd total of nearly 500 such forts in England and Wales: indeed it is but little short of a phenomenon to find Norman castles in the Snowdon range![25]

Earthworks are not always devoid of masonry. Great ditches cut in solid chalk, like those of Old Sarum (Fig. 7), were to all intents and purposes walls, and no one would require further defences:[26] but ditches in earth were often, in stony places, strengthened by a wall on their inner face. In places this wall is very formidable and disposed in the manner of *Iliad*, xii. 258,[27] as at Worlebury in Somerset (Fig. 8), though this seems to have been occupied in pre-Roman times.[28] Where the soil was suitable,[29] the earthworkers often faced their vallum with retaining walls of stone (Fig. 8). Caesar describes the Gauls as building walls of mixed timber and stone:[30] these of course would be too perishable to have survived, though at Murcens, above the valley of the Lot, there is a much-dilapidated camp with timber ties in a dry stone wall.[31]

All that can be said of earthworks in the present state of our knowledge is that they all, almost without exception, show 'concentric' plans. Old Sarum is almost a perfect example of a ring-fortress (Fig. 7), and it has probably no Norman (and certainly no Roman) earthwork in it,[32] while at Maiden Castle in Dorchester (Fig. 6), there are in one part no less than eight lines of defence, one within the other.[33] The gates of earth-camps are often fortified with almost trivial elaboration, as at Dumpton (Fig. 9);[34] and in a great camp in Devon at Hembury near Honiton (Fig. 10) there is at an angle a berm so disposed as to be practically an outwork. The presence of a small ditch across the centre of this camp almost persuades Allcroft to call it a Norman addition.[35] If

[25] Also Ireland etc. [X] [The notion is, of course, less preposterous than Lawrence thought and the number may now be set at over 700. See D. Renn, *Norman Castles in Britain* (London, 1968), 14–17, map D.]

[26] Except the Normans, who did put walls atop. But of course they hadn't all read Allcroft. [R] Except Normans! The result of the recent excavations. [X]

[27] [Trans. A. T. Murray (Loeb Classical Library; London and Cambridge, Mass., 1924), i. 563.]   [28] [On the Iron Age dating, see Hogg, *Hill-Forts of Britain*, pp. 293–5, fig. 104.]

[29] [Or 'made it necessary' or 'advisable'. [R]

[30] [*Gallic War*, vii. 23 (trans. Edwards, 413–15).]

[31] A model of this at St Germain[-en-Laye, Musée des Antiquités Nationales,] a little restored. [X] [See M. A. Cotton, '*Muri Gallici*', in M. Wheeler and K. M. Richardson, *Hill-Forts of Northern France* (Reports of the Research Committee of the Soc. of Antiquaries of London, 19; Oxford, 1957), 159–216.]

[32] In its plan that is: of course the masonry is later. [X] [Note that while the origin of the outer ditch and bank is now considered to be Iron Age, the inner one is certainly Norman: see *Old Sarum, Wiltshire* (HMSO; London, 1970); Hogg, *Hill-Forts of Britain*, p. 257.]

[33] [See n. 22.]

[34] [See *The Victoria History of the County of Devon*, ed. W. Page (London, 1906), i. 582–3.]

[35] [*Earthwork*, pp. 83–6.]

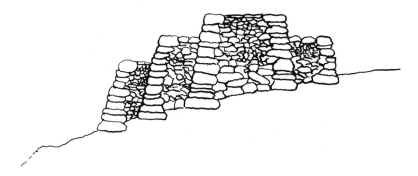

8*a*. Worlbury. Section of wall (Allcroft, p. 176)

8*b*. Cow Castle. Wall section (Allcroft, p. 174)

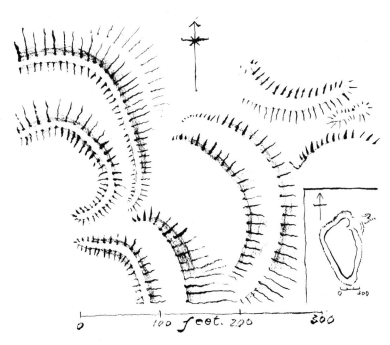

9. Dumpton Great Camp (Allcroft, p. 189)

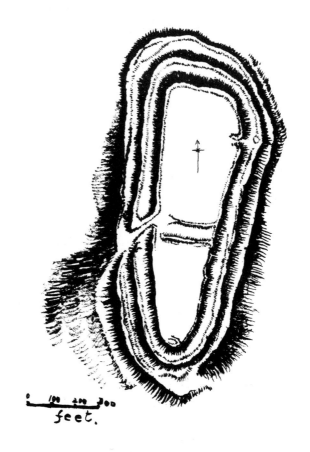

10. Hembury Fort, Devon (Allcroft, p. 85)

this were the case, then the entrance would also be Norman, for the transverse wall is a necessary part of its defence. One may reasonably consider the whole to be pre-Norman, and of one date.[36]

English earthworks are not intended to stand long sieges:[37] only seldom is there water within them. The finest are usually on bare chalk downs, since chalk was so easily worked, and yet so little friable. None the less they are of exceptional interest, as definite forerunners of the multiple castles of the thirteenth century. Their ground plans, for efficiency of defence, have never been improved upon, and they are still extant in such numbers, and often on such a colossal scale, as to give a very high idea of the culture of their engineers. It was in all respects most unfortunate that the clumsy substitute of the imported Norman keep checked their development for nearly a century.

The question as to the amount of skill in mason's work common in Western Europe between the ninth and the eleventh century is not very important.[38] In Paris the inhabitants prepared stone walls and towers against the coming of the Northmen in 886, and a little later Alfred and Ethelfleda repaired the Roman walls of London and Chester.[39] There seems to have been a number of Roman enceintes defensible in England at this period, but no new complete circuits of walls were put up to our knowledge.[40] At the Conquest probably Oxford and Exeter alone of post-Roman towns were stone walled. Exeter is so referred to in 1067,[41] and at Oxford there is still standing a church tower, half-defensive, with signs of communication by bridge with an outer wall.[42] Its date (the date of the wall—not the tower) must be within a few years of the Conquest, certainly before, and also

[36] [In fact the whole fort is now known to be Iron Age, except for the cross-ditch, which belongs to an even earlier, Neolithic phase (see n. 22).]

[37] i.e. they are probably not prehistoric, but defences against raiders: Saxons, Danes, etc. [X] [Many of them are of course pre-Roman.]

[38] i.e. we know nothing about it. [R]

[39] [Paris: see H. Waquet, *Abbon: Le Siège de Paris par les Normands 885–6* (Paris, 1942). London: see n. 17. Chester: see *Anglo-Saxon Chronicle*, trans. Whitelock, p. 61.]

[40] [This statement still appears to hold true (see Radford, *Med. Arch.*, 14 (1970), 102), despite Lawrence's later doubts about it:] Too strong: Saxons were great builders: cf. Malmesbury, which is 9th century in the porches and sculptures. [X]

[41] [Orderic Vitalis, *The Ecclesiastical History*, ed. and trans. M. Chibnall (Oxford, 1969), ii. 210–12. The Roman walls of the late 2nd century had also apparently been repaired by King Athelstan (925–39): see A. Fox, 'Exeter: The Roman City', *Archaeological Journal*, 114 (1957), 178–81.]

[42] [St Michael at the North Gate. Lawrence adds,] Better to put St Michael's down to D'Oigli. [R] [A more recent verdict on the tower, however, is that it is indeed pre-Conquest: H. M. and J. Taylor, *Anglo-Saxon Architecture*, 3 vols. (Cambridge, 1965–78), i. 481–2, figs. 234–5, 546. The town walls, however, would seem to have been of earth and timber before the Conquest, and partly reinforced in stone by Robert d'Oilly soon after: see Royal Commission on Historical Monuments (England), *An Inventory of the Historical Monuments in the City of Oxford* (London, 1939), 159–61.]

Earl Algar in the reign of the Confessor held houses in Oxford to which were attached duties of maintaining the wall in proper repair.[43]

The Normans were not great workers in earth. Normandy has no camp to compare for a moment with Old Sarum: generally her mounds are insignificant,[44] and when William threw up such places in England, as at York, the Saxons found their destruction a mere holiday task.[45] To secure a palisaded camp against fire necessitated a broad berm, and a ditch of many men's labour. The Flemish model after which the Norman nobility had shaped their own castles was a mound with a narrow ring-ditch of such steepness that it had to be crossed by a timbered bridge (cf. Bayeux tapestry). Castle Rising is an English example.[46]

The typical form of castle associated with the Norman is however of course the famous square keep. It is hardly possible to give the Normans credit for its invention. They were not an original race, or rather their originality was shown in the readiness with which they borrowed or adapted the arts, and language, and literature of their neighbours. It would be more natural to find progress in architecture in the tenth century proceeding (e.g.) from Provence, where the little Romanesque keep of Les Baux may quite possibly be older than the keeps of Normandy.[47] Also it has been suggested that in Maine are examples which must be placed before any built by the Normans themselves: a very early keep still exists at Langeais behind the monstrous château of the fifteenth century.[48] There can however be no doubt that the keep took its final shape under the hands of William the Conqueror, whose White Tower in London set the fashion, and became the model for rather more than fifty keeps in England, and nearly as many in northern France.[49] It is perfectly evident that the Tower of London is

[43] Not Algar only. [X]

[44] Arques being an exception. [X] [The castle, of wood and earth (1038–43), was rebuilt in stone in 1123: see R. Quenedey, 'Le château d'Arques', *Congrès archéol. de France*, 89 (1926), 307–18; id., 'Découvertes faites au château d'Arques-la-Bataille en 1928', *Bulletin archéol. du Comité des Travaux historiques* (1930–1), 591–600.]

[45] [On William's 2 castles in York, see Royal Commission on Historical Monuments (England), *An Inventory of the Historical Monuments in the City of York*, ii. *The Defences* (London, 1972), 59–89; P. V. Addyman and J. Priestly, 'Baile Hill, York: A Report on the Institute's Excavations', *Archaeological Journal*, 134 (1977), 115–56.]

[46] [R. A. Brown, *Castle Rising, Norfolk* (HMSO; London, 1978); Renn, *Norman Castles*, pp. 295–8, figs. 63–6, pl. XXXIX.]

[47] [This keep would appear to be 13th century in date: see F. Benoit, *Les Baux*, 3rd edn. (Paris, 1955).]

[48] [Built by Fulk Nerra, Count of Anjou, *c.*994: see A. Blanchet, 'Recherches au donjon de Langeais', *Bulletin monumental* (1931), 74–81; F. Lesueur, 'Le château de Langeais', *Bulletin de la Société nationale des antiquaires de France*, 106 (1948), 378–400 (pp. 378–85); A. Châtelain, *Donjons romans des Pays d'Ouest* (Paris, 1973), 153–4, pl. IX.]

[49] [R. A. Brown and P. E. Curnow, *Tower of London* (HMSO; London, 1984); R. A. Brown,

not the first of its kind; it is too certain in all its details to be an experiment, but the castle at Rouen, from which it is sometimes supposed to have been copied, has conveniently disappeared, and in Normandy the two or three specimens which might possibly be pre-Conquest are very poor and uninteresting.[50] Probably Duke William discouraged the building of keeps by his nobles as far as possible, until the Conquest of England laid open a huge field for the activity of every Norman architect and man-at-arms.

The appearance and arrangement of these keeps are too well known to need illustration. Of course their great principle is passive defence, and to secure it they were built more solidly than almost any building before or since. At Newcastle the lower 14 feet, in a keep 90 feet square, is solid throughout,[51] and elsewhere, though they do not run to this extreme, yet the outer walls and the dividing wall may be anything from 15 to 20 feet thick. The corners are usually strengthened in addition with pilasters, very shallow buttresses in intention. The entrance is always on the first floor: in the early keeps it was often reached by a wooden ladder: in later ones by a fore-building, with at times a drawbridge in it. To prevent all possibility of surprise, the doorway was made very narrow. The parapet is usually plain, there were[52] no hoards and very few loopholes, and the portcullis is nearly unknown. It is obvious that a tower such as this would be impregnable, if mining was impracticable. On the other hand its garrison could only be a scanty one, and once in the keep they could be imprisoned most hopelessly, by a very small force. 'A keep could be defended by one man'—perhaps—but it could certainly be besieged by two, standing one each side of the doorway to prevent egress. There were never any covering works, from which a flanking fire could be maintained, and a sally in any force would be dangerous, owing to the impossibility of retreat in haste.

The Norman keep was thus rather an ineffective fortress; it could be mined with the greatest ease, as King John proved at Rochester,[53] and it

'Some Observations on the Tower of London', *Archaeological Journal*, 136 (1979), 99–108; Renn, *Norman Castles*, pp. 326–30, figs. 70–3, pl. XLIV.]

[50] They are figured in [A.] de Caumont's *Abécédaire [ou rudiment] d'archéologie*[: *Architectures civile et militaire* (Paris, 1853, 3rd edn. Caen, 1869), 408–10. On Rouen and other pre-Conquest castles in Normandy, see R. A. Brown, *English Castles* (London, 1976), 14–39.]

[51] [Not quite. See W. H. Knowles, 'The Castle, Newcastle upon Tyne', *Archaeologia Aeliana*, 4s. 2 (1926), 1–51; B. Harbottle, *The Castle of Newcastle upon Tyne* (Newcastle, 1977); Renn, *Norman Castles*, pp. 254–7, fig. 48, pls. XXX–XXXI.]   [52] Till just at the end. [X]

[53] John didn't attack Rochester apparently; no matter. [R] [King John did indeed besiege the castle in 1215, when his miners succeeded in bringing down the southern corner turret; this was rebuilt subsequently, in 1226–7, with a rounded instead of a rectangular plan. See R. A. Brown, *Rochester Castle, Kent* (HMSO; London, 1969, repr. 1974); id., *English Castles*, pp. 69, 180, fig. 29; Renn, *Norman Castles*, pp. 299–303, figs. 65–6, pl. XL.]

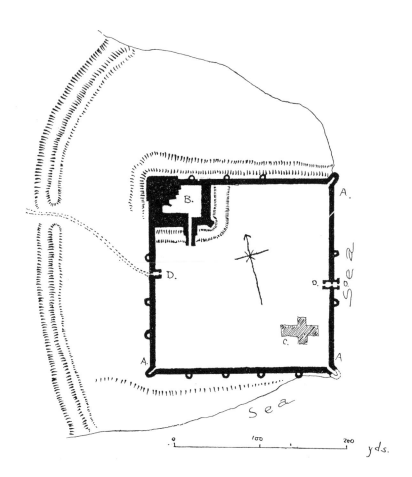

11. Portchester Castle (from Allcroft, p. 414). A–A: Roman enceinte. B: Norman castle. C: Church. D–D: Roman gates. The age of the earthworks across the promontory, round the Roman walls, and round the Norman castle, is not determined. [The earthwork across the promontory is now dated by excavation to the late Saxon or medieval period, possibly as late as 1337: see B. W. Cunliffe, 'Excavations at Portchester Castle, Hants 1963–5: Second Interim Report', *Antiquaries Journal*, 46 (1966), 39–49 (pp. 41–3). Those surrounding the Roman and medieval walls are roughly contemporary with those walls (see n. 54)]

provided little accommodation. Apparently its builders felt its defects, for they never allowed a keep to stand alone. Sometimes it is within a Roman wall, as at Portchester[54] (Fig. 11) and Pevensey,[55] more often there was a ring-wall drawn around it as at Ludlow: and this wall was of course provided with curtain-towers.[56] Other keeps have only earth-works, older or contemporary and once no doubt palisaded, for their outer defence. Usually they stand on the strongest point of the site, but sometimes, as at Richmond, they are so placed as to defend the weakest part of the outer wall.[57] In either case these outworks are seldom elaborate enough to stand the determined attack of a considerable force. The square keep is an ideal defence against a border raid, and in the north of England it survived in this purpose in the peel-towers to the sixteenth century: but against an enemy with leisure, or sufficient resources to drive a mine, the reduction of such a castle was only a question of time. The form probably only remained in favour for a very brief period, but this included the years in which the first Crusaders went out to Palestine. One may determine therefore with reasonable certainty that Normans and Provençals alike were accustomed to build castles of a square keep and a curtain wall with towers around it: only in the Norman castles greater stress was laid on the keep, and in the Provençal on the curtain wall.

IV. BYZANTINE MILITARY ARCHITECTURE

JUSTINIAN'S military architecture had begun in that of old Rome, in imitation of the wonderful system of Roman fortresses in Mauretania.[1]

[54] [See S. Rigold, *Portchester Castle, Hampshire* (HMSO; London, 1965); B. W. Cunliffe, *Excavations at Portchester Castle*, 4 vols. (Reports of the Research Committee of the Society of Antiquaries of London, 32–5; London, 1975–84); Renn, *Norman Castles*, pp. 281–5, figs. 58–9, pl. XXXVI.]

[55] [See Sir Charles Peers, *Pevensey Castle, Sussex* (HMSO; 2nd edn. London, 1953); Renn, *Norman Castles*, pp. 276–9, fig. 57.]

[56] Am not sure about Ludlow, but there are heaps of other examples. [X] [The 12th-century ring-work has rectangular curtain-towers on the side facing away from the town. The keep was established later in the 12th century by transforming an earlier gate-tower, as at Richmond (see below): see W. H. StJ. Hope, 'The Castle of Ludlow', *Archaeologia*, 61 (1908), 257–328; P. E. Curnow, 'Ludlow Castle', *Archaeological Journal*, 138 (1981), 12–14; Renn, *Norman Castles*, pp. 232–3, fig. 40, pls. XXIII–XXVI.]

[57] [The rectangular keep was formed in the late 12th century by remodelling the castle's 11th-century gate-house. See Sir Charles Peers, *Richmond Castle, Yorkshire* (HMSO; London, 1953, repr. 1968); Renn, *Norman Castles*, pp. 294–5, figs. 68–70, pls. XXXVII–XXXVIII.]

[1] [This section is based largely on Diehl, pp. 138–225. For a more recent analysis and criticism of Diehl, see D. Pringle, *The Defence of Byzantine Africa from Justinian to the Arab*

In that province the Romans had had in some way the position of the later Byzantines: ordinarily their defensive works were slight—little more than walled *castra*, along a frontier—but in Mauretania they were a small garrison holding a difficult country against an enemy that attacked continuously and unexpectedly, though without any very great knowledge of war. So the Roman forts there were erected on the best strategical positions, or across the great military roads, or next to some town to provide a refuge against sudden incursions: and when the Byzantines under Justinian fortified their empire systematically, they followed the same plan. Their buildings may be divided into fortified towns, refuge camps, castles properly so called, single towers to defend isolated farms, and signal posts. A close network of these was erected up the Euphrates and in Asia Minor, on the Danube, and throughout northern Africa, hundreds of them in the comparatively short period of the reign of Justinian. After him Byzantine architecture stood almost still: no one has yet been able to distinguish later Byzantine repairs and additions amongst the original works of some 'Justinian' castle:[2] the Greeks were so obsessed with the excellences of his work, so bound by the precepts of Procopius and of the author of the *Tactica* who followed him, that they made hardly any improvements. The great fortresses which the Crusaders came upon in their march, Nicaea, Antioch, Edessa, are standing today, perfect examples of the style developed by the Emperor, and the architects whom he instructed.[3]

Constantinople of course was in theory earlier;[4] but it shows how nearly allied was Justinian's work with that of his predecessors. There is, however, very little variation in the Emperor's plans. He was a pedant, and laid down for his fortresses exact rules which his

Conquest, 2 vols. (British Archaeological Reports, International Series, 99; Oxford, 1981), i. 94–109, 131–66.]

[2] No one has tried to. In any case the statement only refers to the plan: the masonry became steadily worse with time. [X]

[3] They are pulling down Edessa. [X] [The walls of Nicaea date from the time of Claudius Gothicus, AD 268–9: see A. M. Schneider and W. Karnapp, *Die Stadtmauer von Iznik (Nicaea)* (Istanbuler Forschungen, 9; Berlin, 1938); A. M. Schneider, 'The City Walls of Nicaea', *Antiquity*, 12 (1938), 437–43. The fullest account of the defences of Antioch is still Rey, pp. 183–204, pl. XVII. Recounting his visit to Edessa in Sept. 1909, Lawrence wrote, 'I found there the only two beaked towers in all N. Syria. I hope my photos will be clear. They were the last I took' (*Home Letters*, p. 108). His second visit, in 1911, is described below, pp. 137–8. On the defences of Edessa, see H. Hellenkemper, *Burgen der Kreuzritterzeit in der Grafschaft Edessa und im Königreich Kleinarmenien* (Bonn, 1976), 31–7, pls. 1–5, 73–4. These and other examples of Byzantine fortification are discussed by A. W. Lawrence, in 'A Skeletal History of Byzantine Fortification', *Annual of the British School at Athens*, 78 (1983), 171–227, pls. 8–21.]

[4] Probably in practice also! [R] [Its walls date from the reign of Theodosius I, AD 412–22. See *Die Landmauer von Konstantinopel*, vol. i by F. Krischen (Berlin, 1938), vol. ii by B. Meyer-Plath and A. M. Schneider (Berlin, 1943); A. M. Schneider, 'The City Walls of Istanbul', *Antiquity*, 11 (1937), 461–8.]

lieutenants were generally unable to refuse, except in cases of exceptional haste or on some unimportant occasion. Save for such buildings, we find Procopius' instructions excellently well illustrated in Greek military engineering. First of all it departed from the Roman system of a line of camps at intervals along a frontier, as bases for a vigorous offensive. Justinian ordered a line of small strongholds along the frontier, much more closely placed than were the Roman camps, but of less size, and weaker quality.[5] Their efficiency deterred the private raider, the artist on a small scale, sufficiently well. Against a strong attack they were only to serve as signal posts, and as outposts, to delay the enemy a day or two, until the people of the province and their goods had been gathered within the second line of defence. This was composed of central citadels, erected as refuges wherever population warranted. The army likewise stood on the defensive until these refuges were occupied: and then the enemy, distracted by the number of their potential sieges, and the hollowness of their occupation of the country, were easily driven out: more especially as by their fire-signals the Greeks were able to exchange information, or to concentrate at a speed that was disconcerting to the most active of invaders.

The buildings themselves, according to Procopius,[6] were to have three lines of defence. First would be the ἀντιτείχισμα, a mound piled up of the earth taken out when cutting the ditch within it. The ditch was to be not less than 56 feet wide,[7] and as deep as the foundation of the inner walls to discourage sap. It was an advantage if it could be water-filled; in any case its sides had to be perpendicular. The mound outside was to increase its apparent depth, to hide it, and to force assailants to expose themselves to an easy fire from the wall.[8]

Within the ditch was the προτείχισμα: a wall of some height, banked up with earth within, so that its *chemin de ronde* would be nearly on a level with the lists. In these περίβολοι [walls] the country people would be assembled, and their defence would naturally be the more strenuous, since they would know the exceeding unlikelihood of rescue from the defenders of the inmost wall. Procopius lays down that the space between the two walls should be one-quarter the height of

[5] Procopius, *Buildings*, iv. 1, 33–6 [ed. J. Haury (Leipzig, 1913), p. 68 (trans. H. B. Dewing (Loeb Classical Library; London and Cambridge, Mass., 1940), p. 229); cf. Diehl, p. 143.]

[6] [This advice is given not by Procopius, who did no more than record the building achievements of Justinian, but by the anonymous author of the *Tactica* (or *de Re Strategica*), ed. H. Köchly and W. Rustow, *Griechische Kriegsschriftsteller*, ii. 2 (Leipzig, 1855), ch. xii. Lawrence's mistake was evidently derived from Diehl, p. 145.]

[7] [i.e. 40 cubits: cf. *Tactica*, xii. 6.]

[8] This is my reason for a mound, not Procop[ius']. [R] [The outer bank is mentioned by Diehl, p. 146, but not by Procopius or the treatise; the latter advises only that material from the ditch be spread between the 2 walls; but if the defences were on a hill slope, then an outer bank and scarp could be substituted for the outer wall and ditch (xii. 6–9).]

11a. Enfeh (Nephin): the moat looking north. The mass of rock in the centre is for a bridge-pier: this, like Athlit, is a sea-level cutting. [August 1909. See Deschamps, *Châteaux*, iii. 297–301; pls. LXI–LXIII]

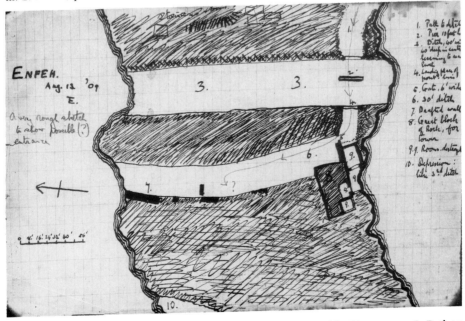

11b. Enfeh, 13 August [19]09. A very rough sketch to show possible (?) entrance. 1: Path to ditch. 2: Pier 10 feet high. 3: Ditch, 40 feet wide, 40 feet deep in centre, lessening to sea level. 4: Landing place of pont-levis. 5: Gate, 6 feet wide. 6: 30 foot ditch. 7: Drafted wall. 8: Great block of rock, for tower. 9.9: Rooms, destroyed. 10: Depression: like third ditch

the inner:[9] and this unusual nearness was intended to enable both walls to be manned at once against attack. The inner wall, τεῖχος, was to have two rows of defences: the ground floor of loopholes in embrasures of some size: the upper a gallery, often vaulted, or wood-roofed, sometimes open, shielded outside by merlons of a certain strength and height. The thickness of a wall should be one-fourth its height and at intervals upon it must be towers, three-storied, and usually square.

The Byzantine curtain-towers are mysteriously inadequate. The *Tactica* demands that towers be octagonal outside, and circular inside:[10] and one of this pattern exists at Bash Dagh in Asia Minor: but generally there are none such.[11] Occasionally towers are octagonal, inside and out, more often hexagonal: nine-tenths are simply rectangular: in the African fortresses round towers are sometimes used at the angles of the larger places, and very occasionally there is one on the curtain wall. Procopius mentions towers that commence square, with a circular superstructure;[12] and there is one at Bash Dagh,[13] and three or four in Africa. The shallow rectangular shape is however the usual one: the towers have the thinnest walls on the enceinte, often only half the thickness of the curtain, and are hollow to the ground level. There is hardly a Byzantine tower that could not be smashed in with a few blows of a ram.[14] Their square fronts made attack easy, and mangonel stones found a fair target: also the square shape gave very little flanking fire (Fig. 12), and was less defensible from the walls.[15] Before an earth-quake it was most liable to collapse. The only point in its favour[16] was its readiness of construction; that they valued more the round tower, or the polygonal was shown by their placing these at important points (Fig. 13). For the rest they seem to have trusted to the weakness in siege craft of the enemies they had to ward against: the Arabs were till Saladin's time contemptible engineers: and the Greeks found that a plain wall without towers was often sufficient to check them. Curtain-towers were only seldom (as at Antioch) connected both with the

[9] [Again, the author of this statement is not Procopius, but, in this case, Diehl, p. 145.]
[10] [In practice the description given in the text (xii. 2) indicates a 4-sided projecting tower shaped like a cut-water (the type referred to by Lawrence as 'beaked'). For 6th-century examples see Pringle, *Defence of Byzantine Africa*, i. 157–8; and A. W. Lawrence, article cited in n. 3.]
[11] Constantinople has lots. [X] [but they do not have beaked fronts: (see n. 4). At Bash Dagh there were at least two polygonal towers on rounded bases with rounded inside walls: see W. M. Ramsay and G. L. Bell, *The Thousand and One Churches* (London, 1909), 285, figs. 239b, 241–3.]
[12] [*Buildings*, ii. 1, 18–19, ed. Haury, p. 30 (trans. Dewing, p. 107).]
[13] [Apparently not: see n. 11.]
[14] Not having used a ram myself. [R]
[15] Being of small projection, with dead centre. [R] All this from Mr Lethaby. [X]
[16] According to Prof. [W. R.] Lethaby. [R]

12. 1. Rectangular curtain-tower
    2. Half-round tower of the same diameter: showing its advantages for defence and offence

13. Polygonal tower at Edessa, at angle. From across moat. [September 1909.] (Second visit 1911: climbed this tower, and found certain signs of *Arab* work in it: at the same time the argument may stand, for the foundation is earlier work)

13a. [Edessa. Beaked tower. September 1909]

chemin de ronde and the interior of the fortress. More often the towers are isolated from the curtain; sometimes there is no entrance from the interior: when there is there is usually no communication with the upper floor: and the entrance was always inconveniently narrow, sometimes less than two feet wide, in a passage of ten feet. They were always stone-vaulted: probably simply from lack of wood.[17]

We hear of one other part of a castle, the φρούριον or πυργοκάστελλον of Procopius.[18] It took the place of the Western keep: and like it was usually on the wall of the building, and at one end: only, while the Frank donjon was usually put in the strongest defensive position, the Byzantine one was sometimes put in the weakest. The more common position, however, was just as in the West, though of course nothing so defenceless as the Norman keep was employed (Fig. 14).

The gates of their castles were the chief concern of the Byzantine architects.[19] They had no appreciation of the portcullis or herse:[20] and so their gates are more complicated than the Latin ones: also the Arabs seem to have attacked, to some extent, the gates in particular. At any rate they were made extremely narrow (quite a large one is only 4 feet wide at one point) and whenever possible were flanked by towers at the sides. Sometimes they were set in towers, and defended by being forced around at right angles in a trap as at Ain Tounga (Fig. 15).[21]

At Mdaourouch (Fig. 16) they were of a different form: the double gate here may have been intended to surprise a small attacking party. A force might be hidden in the side chambers, and overlooked by a party battering the inner gate; but more probably it was only for the convenience of guards in the time of peace.[22] It could be arranged that one gate should be shut before the other was opened: and so surprise would be made impossible. At Timgad (Fig. 17) these two systems were combined into one extremely formidable entry.[23]

---

[17] [This assessment is based largely on Diehl, pp. 156–8. In fact, 6th-century towers were usually accessible both from the wall-walk and from within the enceinte at ground level: see Pringle, *Defence of Byzantine Africa*, i. 152–8.]

[18] [*Buildings*, ii. 5, 8–9, ed. Haury, p. 38 (trans. Dewing, p. 135); cf. iii. 5, 11, iv. 11, 16, ed. Haury, pp. 59, 90–1 (trans. Dewing, pp. 205, 307). For discussion, see Pringle, *Defence of Byzantine Africa*, i. 155–6.]

[19] In Africa. [R]

[20] [On the Byzantine use of the portcullis, see Pringle, *Defence of Byzantine Africa*, i. 160–2.]

[21] Diehl, pp. 159–60, figs. 12–13. [Cf. Pringle, *Defence of Byzantine Africa*, i. 162–3, 271–2, pl. LXXIIa.]

[22] [Diehl, p. 205, fig. 14. There is no evidence for the side chambers shown on Diehl's plan (here Fig. 16): see Pringle, *Defence of Byzantine Africa*, i. 161, 216, figs. 11–12, pl. XIXb.]

[23] [Again, Diehl's plan (fig. 37: here Fig. 17) is misleading: the outer bent entrance is of a much later date, and the single gallery (in only one of the side walls, above the heads of would-be attackers) was probably intended as a command post for controlling the operation of the portcullis, which represented the outer of the two systems for closing the gate-passage. See J. Lassus, *La Forteresse byzantine de Thamugadi* (Paris, 1981), i. 78–89, figs. 45–54; Pringle, *Defence of Byzantine Africa*, i. 161, 234–5, figs. 2, 6, pls. XXXVa, XXXVIa–b.]

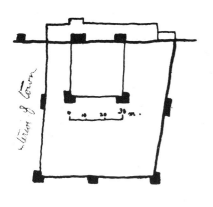

14. Kasr Bagai. Byzantine
πυργοκάστελλον (Diehl, p. 193)

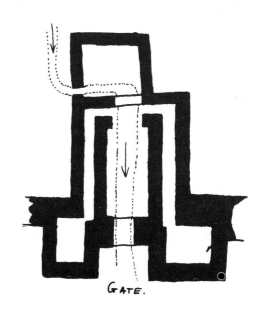

16. Mdaourouch. Gate (Diehl,
p. 161)

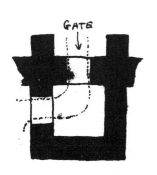

15. Aïn Tounga. Gate (Diehl,
p. 159)

17. Timgad. Gate (Diehl, [p. 203])

Of other details it is worth noting that the *chemin de ronde* had frequently to be carried on corbelling or on arcades, owing to the slightness of the walls, as at Antioch and elsewhere: and that no *machicoulis* of any sort or pattern appear on any known Byzantine fortress prior to the thirteenth century.[24] One can only conclude that they were unknown.

In plan Byzantine castles are found of two kinds: one a derivation of Roman camps, square with towers at the angles, and if the curtain is over-long, wall-towers as well at rare intervals.[25] The other is the fortress built on some easily defended position, following in its lines the contour of the ground. Generally, however, the Byzantine fortresses are not skilfully placed, as regards defence: they are frequently over-looked in an extraordinary manner from hills close by: and they never preferred a defensible post to one strategically important as the Crusaders so often did. The Greeks put their buildings where they were wanted: the Franks where they would be impregnable. The Greek forts were defended by the size of their garrisons, the depth of their ditches, and the efficiency of their army in the field; the Latin fortresses by their carefully schemed covering fire, and the natural advantages of their position.

V. THE ARCHITECTURE
OF THE CRUSADERS IN SYRIA

IN treating of the Latin fortresses in Syria itself, it cannot be too strongly urged that documentary evidence of building is absolutely valueless. Actual examination of the existing ruins often shows that the Crusaders only occupied (or at the most repaired) an already existing Byzantine castle: or there may be signs of later rebuilding, either by a succeeding generation of Franks, or by Beibars,[1] who was frequently generous enough to extend his aegis over the works of his predecessors, by cutting in his own inscriptions; or even, as in the Sidon district, by Arabs of the sixteenth or seventeenth centuries.[2] Medieval fortresses must in every case be dated from their own evidence.

---

[24] One is on the 'Constantine' gate in Constantinople. I think late. [X] [Lawrence visited Istanbul in Dec. 1910: see *Home Letters*, pp. 116–20. A number of 6th-century examples are also known: see Pringle, *Defence of Byzantine Africa*, i. 165–6 and notes.]

[25] At Constantinople 180 ft. apart in places.

[1] [Baybars I, Mamluk Sultan, 1260–77.]

[2] Fakhreddin. [R] [i.e. Fakhr al-Dīn II, Druse amir who controlled most of Lebanon and northern Palestine, c.1590–1633.]

In dealing with the twelfth century in the East, Arab influence in architecture may be entirely discounted.[3] Beibars seems to have been the first Arab sovereign to build respectable fortresses. His citadels in Aleppo (Fig. 17a) and Damascus, and his work at Crac des Chevaliers are creditable, if not very remarkable copies of Greek or Latin work.[4] Before him the masterpieces of Syrian unaided effort are to be seen in the absurdly weak castles at Masyad and Kadmus.[5]

Any very great uniformity in the Syrian fortresses of unquestioned Latin parentage of the twelfth century would be surprising. The piecemeal character of the Frankish conquest of the country, and the fact that their political divisions often implied racial differences, would tend towards the formation of local schools of architecture and, as a matter of fact, castles quite opposed in principle were erected at the same time in different districts. The Normans of Sicily had adopted many Eastern customs: in Antioch and Edessa they adopted the Greek fortresses (Fig. 18): a few slight walls in the castle at Antioch are the only sign of Latin occupation in that province.[6] In Tripoli great builders among the nobles were succeeded in their fiefs by the Military Orders, who carried out—or surpassed—the intentions of their predecessors, and accordingly we find there the most elaborate fortresses. In the southern part of Syria, that depending particularly on Jerusalem, very few castles have been preserved. Later occupation is responsible for the

[3] The Arabs had Sassanian models in palaces, but required no new fortresses: they were offensive, not defensive: and had no need of a military architecture. [R] [See, however, K. A. C. Creswell, 'Fortification in Islam before A.D. 1250', Proceedings of the British Academy, 38 (1952), 89–125, pls. 1–16; A. Lézine, Le Ribat de Sousse suivie de notes sur le ribat de Monastir (Notes et Documents, 16; Tunis, 1956).]

[4] [At Aleppo (see Fig. 17a), the stone glacis and massive gate-tower, including 6 right-angled turns in the gate-passage, were built by Malik al-Ẓāhir in 1203/4: cf. Creswell, Proc. Brit. Acad., 38 (1952), 124–5, fig. 16, pl. 16; S. Saouf, La Citadelle d'Alep (4th edn. Aleppo, 1975); M. Haraytānī and S. Shaʿash, Qalʿat Hālib (Aleppo, n.d.). In Damascus, the Citadel was founded by Malik al-ʿĀdil in 1208; though additions were made later by Baybars amongst others: see D. J. Cathcart King, 'The Defences of the Citadel of Damascus: A Great Mohammedan Fortress of the Time of the Crusades', Archaeologia, 94 (1951), 57–96, pls. xvii–xx. Crac des Chevaliers is described below.]

[5] The account of Masyad by Rider Haggard in The Brethren is splendidly imaginative. The genuine place is contemptible. (It serves as the workhouse of its district. [R]) [On Masyaf, see P. Deschamps, Les Châteaux des croisés en Terre Sainte, iii. La Défense du Comté de Tripoli et de la Principauté d'Antioche (Bibliothèque archéologique et historique, 90; Paris, 1973), 39, pl. xcii; and W. Müller-Wiener, Castles of the Crusaders (London, 1966), 68–9, pl. 95. The castle at Qadmus has been destroyed, though its site, on an outcrop of rock, is still clear enough; it is occupied by a village, in which Lawrence spent a night in Aug. 1909 (Home Letters, p. 105). Another castle of this group, Msaylha, is illustrated in Fig. 33a.]

[6] [On Antioch and Edessa, see p. 26 n. 3. Knowledge of Frankish works in the Principality of Antioch is somewhat greater now than it was when Lawrence wrote: see Deschamps, Châteaux, iii. The principal elements of the massive castle of Baghras may also now be added to the catalogue: see R. W. Edwards, 'Baǧras and Armenian Cilicia: A Reassessment', Revue des études arméniennes, NS 17 (1983), 415–55, pls. lx–lxxxii; cf. A. W. Lawrence, 'The Castle of Baghras', in T. S. R. Boase (ed.), The Cilician Kingdom of Armenia (Edinburgh, 1978), 35–49.]

17a. The viaduct at Haleb [Aleppo]. I think that all the fortress follows Byzantine lines: except perhaps this thing: which is extraordinary. [August–September 1909]

destruction of most, but there are some, beyond the Dead Sea, still standing tolerably perfect, but quite undescribed. Rey has published a plan and description of Kerak in the Desert, but neither is his own work:[7] indeed Kerak has never been studied by a medievalist.[8] It has been hailed by Rey, and by Professor Oman following him, as an untouched example of Latin military architecture. This claim seems a little dangerous, when it is remembered that Kerak was a Byzantine fortress before it became Crusader, and that after this it was the seat of a powerful Arab principality, and that finally Beibars's presence is shown by his name on one of the towers. It may well be that the share of Payn of Nablous in the building of it is infinitesimal.[9] At least, until there is better material to work upon, elaborate deductions from it as to the state of Latin military architecture in 1140 are quite out of place. Beibars's buildings in particular are very easily confused with Crusader work, for it seems most likely that they are a partial imitation.[10]

To consider the Crusading castles in their chronological order is extremely difficult: they are mainly a series of exceptions to some undiscoverable rule. To begin with the castles of the Antioch and Edessa principalities means beginning with a string of nearly untouched Byzantine fortresses. At Antioch the walls have no signs of Latin interference: and the castle (Fig. 18), with its long, flimsy wall with the ridiculous buttresses, is evidently residential and not a post of military importance.[11] In the Antioch province the celebrated castle of Harenc (Harim) is Byzantine and Arab. It has a gigantic dry ditch cut in the rock on which the castle stands, and this alone would be sufficient proof of a Greek origin: when the case is supported by a ground-plan (Fig. 19) so typically Byzantine there can be no possible hesitation in dating it before the arrival of the Crusaders: very little of its walls remain.[12]

---

[7] [*Étude*, pp. 132–5, pl. xiv.]

[8] And the unthinking activity of some Bedawin, in tearing up the Hedjaz railway near Amman, prevented my going there in 1909. (And in 1911: only this year the Druzes helped the Arabs. They have killed the Turkish governor, and burnt the konak and the schoolmaster. [X])

[9] From photographs Kerak seems fairly late in date: perhaps ? Reginald of Châtillon. [X]

[10] [The investigations of P. Deschamps have gone some way to distinguishing the parts of the castle built by Pagan the Butler in 1142 from those of Sultan Baybars and others in the 13th century: see *Les Châteaux des croisés en Terre Sainte*, ii. *La Défense du royaume de Jérusalem* (Bibl. archéol. et hist., 34; Paris, 1938), 80–98, pls. iv–xxvii, plans; cf. Müller-Wiener, *Castles*, pp. 47–8, pls. 23–7. Twelfth-century work seems to be confined to the inner ward and there is no Byzantine work apparent; further clearance is needed, however, to complete its investigation.]

[11] It's on a mountain peak and very hard to get at. [X] [See p. 26 n. 3. The Citadel, however, does appear to contain Frankish work.]

[12] [In a letter of Aug. 1909, Lawrence writes, 'At Harim I found a Crusading castle, too ruined and rebuilt to be valuable, with underground passages that would have rejoiced the rest of the world: I explored them with candles . . . The passages went to water-springs etc.'

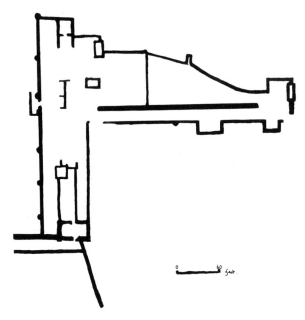

18. Antioch. The castle (C. H. C. Pirie-Gordon)

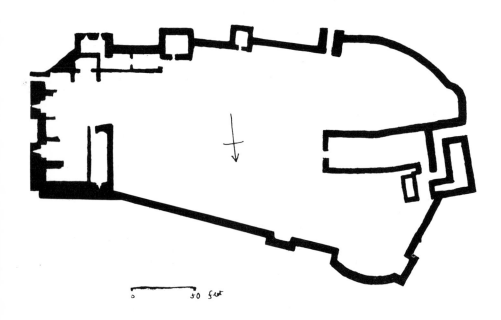

19. Harim (C. H. C. Pirie-Gordon)

Going further east, the site of Turbessel (Tell Bashar) is entirely laid waste. There had been once a little castle on a mound, but only a few stones of a square tower are left.[13] At Bira (Biredjik) on the Euphrates was a fortress (Fig. 20) which would be typically Byzantine if the Arabs under Malek es Zahir had not rebuilt the high towers looking south-ward:[14] and the huge castle of Edessa (Urfa, Rages) is also Byzantine with Arab additions. The rock moat (Fig. 21), over 500 feet long, with an average depth of 60 feet, and a width of 30 feet is too huge a work for the Latins ever to have undertaken in their insecure tenure of the place: besides we have there town walls tolerably perfect,[15] and unquestion-ably Byzantine, and the Crusaders appear to have maintained a semi-Byzantine administration during the few years they held the province. It is important, having regard to other castles in northern Syria, to notice the pier[16] of rock (Fig. 21) left standing in the moat when the rest was excavated: its purpose was to support the centre of a timber bridge, and to make it high enough it was necessary to cap it with masonry: the only large wood to be found in Edessa today is the poplar tree, and a long beam of this is quite untrustworthy, even in the far less trying strain of the roof of a native hut. In Europe a drawbridge could be made quite long without any sacrifice of stability, and therefore the moat pier was not required. One example however is to be found at Tonquédec in Brittany, in work of the fourteenth century.[17]

In Rum Kalaat [Fig. 21a], another stronghold of Edessa, the only sign of Latin occupation is in the form of the grooves for a portcullis.[18] For some reason the Byzantines never used this defence for a gateway with

(*Letters*, p. 78; *Home Letters*, p. 106). Although a Byzantine castle stood at Ḥārim from 959, the existing remains appear to be largely those of the rebuilding, carried out by Malik al-Ẓāhir, Sultan of Aleppo, in 1199. The Franks only held the castle from 1098 to 1149 and from 1158 to 1164, and the extent of their contribution to its defences has yet to be assessed. See M. Van Berchem and E. Fatio, *Voyage en Syrie*, 2 vols. (Mémoires de l'Institut français d'archéologie orientale du Caire, 37–8, Cairo, 1914–15), i. 229–38, figs. 139–40; Deschamps, *Châteaux*, iii. 341, pl. LXXIIIb; Müller-Wiener, *Castles*, p. 65, pls. 88–9.]

[13]   [See Hellenkemper, *Burgen der Kreuzritterzeit*, 38–43, pls. 6, 75; H. P. Eydoux, 'Le château franc de Turbessel', *Bulletin monumental*, 139 (1981), 229–32.]

[14]   [See below, p. 142.]

[15]   Town walls now pulled down. [X]

[16]   Two piers. [X]

[17]   [Or early 15th century: see A. de la Barre de Nanteuil, 'Le Château de Tonquédec', *Bulletin monumental*, 75 (1911), 43–75. Lawrence visited and described this castle at length in Aug. 1906 (*Home Letters*, pp. 16–20).] Another at Caen, 14th, and one at Coucy, ?13th. [X] [The drawbridge at Caen rested on a pinnacle of masonry in the centre of the moat: see M. de Bouard, 'Le Château de Caen', *Congrès archéol. de France*, 132 (1974), 9–21 (p. 17). On Coucy, see E. Lefèvre-Pontalis, 'Coucy-le-Château', *Congrès archéol. de France*, 78. 1 (1911), 293–308 (p. 306); id., *Le Château de Coucy* (Paris, 1913).]

[18]   There are no portcullises in Rum Kalaat. It is all medieval Armenian and Arab. [X] [Lawrence's description of this castle in July 1911 is printed below, pp. 142–7. See also Hellenkemper, *Burgen der Kreuzritterzeit*, pp. 51–61, pls. 9–11, 89.]

20. Bira (Biredjik) from the river. [September 1909]

21. The moat at Edessa with drawbridge pier.
[September 1909]

21*a*. [Rum Kalaat. The west front of the castle, including one of the
gate-towers, looking north. Taken in September 1909]

any frequency. They knew of it of course, for it is described by Vegetius, and found in action in Pompeii and the Great Pyramid;[19] and judging from their practice European castle-builders found them profitable. In the East they may be taken invariably as tokens of European influence.

These castles are a little disappointing, but there is one great castle, depending on Antioch, that of Saone (Sahyun) which, taken as a whole, is probably the finest example of military architecture in Syria.[20] Like the other castles in the north the remains are very largely of the Byzantine period, but added to in a very definite and very evident fashion by the Crusaders. The castle is built on a narrow ridge (Fig. 22), isolated from its neighbours on each side by being placed in the sharp angle of two streams just about to meet. The valleys on this side and on that are extremely narrow, and some 400 feet deep.[21] Across one of them half-way up had been thrown a bridge, leading to a road cut in the cliff face over what in winter is[22] an impassable torrent. From the bridge the way into the castle leads through a rock moat isolating the end of the promontory, and along the further side of the ridge, to a gate nearly midway in the southern face of the fortress (Figs. 22 and 23). It is unfortunate that the section of this gate in Rey is entirely misleading. There is no portcullis, and no *machicoulis* above, but the doorway is sunk a few inches within the face of the wall, and this depression carried up to a height of some 30 feet, in a shallow blind arch.[23] The plan of the castle given by him, and repeated here (Fig. 22) with the more obvious corrections, is also apparently hurried: Saone is of such colossal size, and so deeply set in inhospitable hills that a complete examination of it is a matter of some exertion and discomfort.[24] Rey's sketch of the great rock moat minimises its very striking proportions; and his description is vague and inadequate. The Byzantine work, which includes all but the great square keep marked E (Fig. 22) is in plan just like any ordinary

[19] Two typical Byzantine constructions! [R] [Vegetius, *Epitoma Rei Militaris*, iv. 4 (cf. trans. in Creswell, *Proc. Brit. Acad.*, 38 (1952), 111).]

[20] [Lawrence visited the castle in late Aug., early Sept. 1909: 'Sahyun, perhaps the finest castle I have seen in Syria: a splendid keep, of Semi-Norman style, perfect in all respects: towers galore: chapels, a bath (Arabian) and a Mosque: gates most original: and a rock-moat 50 feet across in one part, 90 feet in another, varying from 60–130 feet deep: there is a cutting for you! And in the centre had been left a slender needle of rock, to carry the middle of a drawbridge: it was I think the most sensational thing in castle-building I have seen: the hugely solid keep upstanding on the edge of the gigantic fosse. I wish I was a real artist. There were hundreds of other points of interest in the buildings. I stayed there two days, with the Governor . . .' (*Letters*, pp. 77–8; *Home Letters*, p. 106).]

[21] Quite precipitous too. [X] Saladin threw stones into the castle from across the valleys. [R] [23–4 July 1188. On the siege, see Deschamps, *Châteaux*, iii. 229–30.]

[22] Presumably! [R]

[23] [Rey's fig. 34 is certainly incorrect, but the gate *is* defended by a machicolation: see Deschamps, *Châteaux*, iii. 238, plan 7; and id., 'Les entrées des châteaux des croisés en Syrie et leurs défenses', *Syria*, 13 (1932), 369–87 (pp. 377–8, pl. LXXXI).]

[24] I had malaria rather heavy those days. [X]

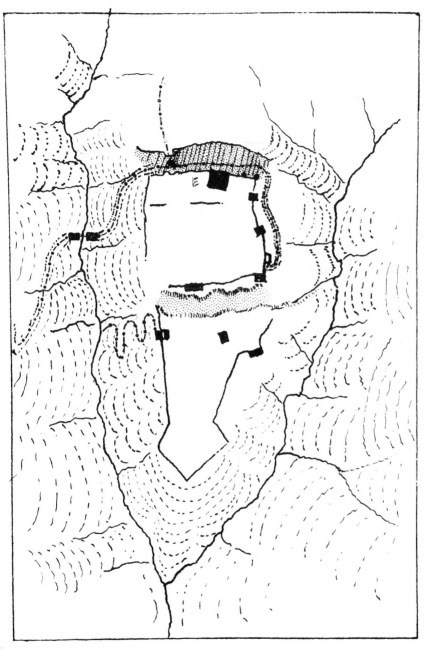

·—·—·—·· Entrances   Rock-cut Ditch

22. Saone (Sahyun) (Rey, p. 107). [E=donjon]

23. Sahyun. The south-east corner. The tower of entrance is the furthest to the left. The great moat runs along before the round tower on the right

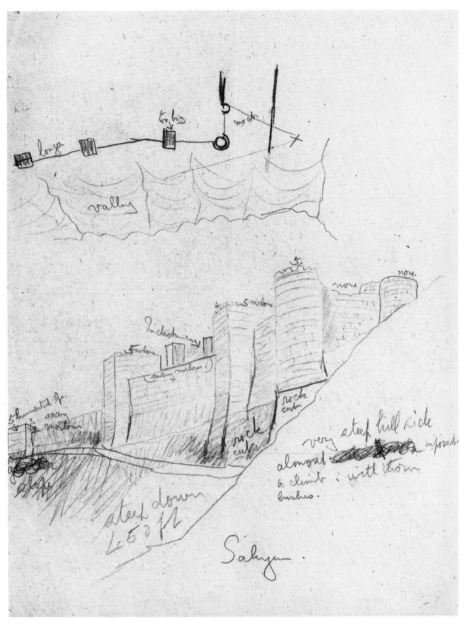

23*a*. [Ṣahyūn. Pencil sketch made at the site, September 1909, probably based on Rey, fig. 31]

Byzantine work, though of very exceptional quality. The moat in particular is in places over 100 feet in width, and the pinnacle to support the bridge stands 110 feet high with its cap of masonry (Figs. 24 and 28). The whole moat is very beautifully carved out of the rock, and its sides have generally been finished to a comparatively smooth surface. It separated the castle from the broader part of the ridge, the part that runs up into the Jebel Daryous.[25]

The keep of the Latin owners of Saone is however of more particular interest. One of a fairly numerous class in Syria, it bears a distinct resemblance to the keeps of north-west Europe, only modified to suit the local conditions (Fig. 25). In Europe no keep was vaulted above the basement; in Syria there was no other way of making a roof.[26] When so large a building was stone-vaulted, however, the height of the European keeps became impossible: the architect of Coucy could build a vaulted keep of four stories, but not so the architect of the early twelfth century.[27] With the abolition of the upper stories went naturally the entrance by ladder or fore-building on the first floor.

Square keeps are to be found in Syria at Saone in Antioch, at Chastel Blanc, Chastel Rouge, Botron and Giblet in Tripoli and at Beaufort, Banias, and possibly Caesarea in the Kingdom. This one at Saone is the most massive (Figs. 25, 26, 27): in form it is roughly a square of about 90 feet each way: the height is 76 feet, and the thickness of the walls in the first floor some 22 feet.[28] The entrance is on the ground level by a very small doorway, closed only by a hinged gate. The keep at Giblet, probably the latest in date, is the only one that has a portcullis of those in Syria:[29] and in Europe also a portcullis in a keep is a rarity and a sign of late date. The gateway gives on a flight of steps, leading in the

[25] [For a detailed description and analysis of the castle see now Deschamps, *Châteaux*, iii. 217–47, pls. VIII–XXX, plans 1–7; and the same author's 'Le Château de Saone', *Gazette des Beaux-Arts* (Dec. 1930), 329–64. Other details will be found in Van Berchem and Fatio, *Voyage*, pp. 267–83, pls. LIX–LXII; G. Saadé, *Château de Saladin (Qalaat Salah-ed-Din)* (Publns. Almanara; Lattakia, 1966); Müller-Wiener, *Castles*, pp. 44–5, pls. 12–21. Note that Lawrence's plan (Fig. 22), copied from Rey (pp. 105–13, fig. 32), foreshortens considerably the length of the castle, which measures in reality some 150 by 740 yards in maximum overall dimensions, the walls following the edge of the natural spur.]

[26] No wood. [X]

[27] He could have but didn't want to. Coucy was a bit of swank on Enguerrand's part. [R] [The cylindrical keep at Coucy was completed by its lord, Enguerrand III, in 1230 (see n. 17). The keep at Ṣahyūn is dated 1108–32.]

[28] There is a lusty colony of snakes in the ground floor, preventing exact measurement; above all since it is in total darkness. (No puttees![X]) [Deschamps gives the tower's measurements as 24.5 m. (80 ft.) square, with walls 4.4–5.4 m. (14 ft. 4 ins.–17 ft. 8 ins.) thick: see *Châteaux*, iii. 239–40, plan 4.]

[29] [Both Jubayl and Ṣahyūn have portcullises defending the entrances to their keeps (cf. Deschamps, *Châteaux*, iii. 211, 239, plans). In the former the portcullis is preceded by a *machicoulis*. At Ṣāfīthā there is no portcullis, but only a *machicoulis* in the ceiling just inside the door (Rey, figs. 26–7).]

24. The moat at Saone

25. Keep of Saone from south-west [September 1909]

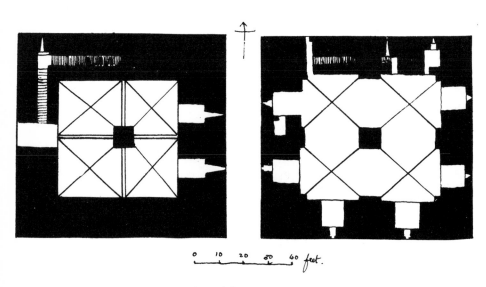

0   10   20   30   40 feet.

26–27. Saone (Sahyun). [Plan of donjon at ground- and first-floor levels]

thickness of the wall to the first floor. Each storey is stone-vaulted, on one huge pier, in the centre, 9 feet square. The upper room is well lighted, by windows of reasonable size, and has garderobes and with-drawing rooms, like any keep in Normandy.[30] A staircase in one corner leads in the wall to the roof, which is surrounded by a parapet on low arches, pierced with loops. This parapet is Byzantine in design, though if, as seems probable, some lord of Sicily drew the plan of the keep, he may have copied buildings standing in his home: in other words this parapet of the keep need not necessarily be[31] a copy of the parapet of the Byzantine curtain wall just below. The straight staircases, and the drafted blocks of which the tower is built are of course not European features. The Crusaders brought with them to Syria their architects, who also acted as chief masons: but the mass of the work must have been done by the natives of the country, the Syrians accustomed to build Greek fortresses. They naturally adopted their own technique in doorways and staircases, and ways of dressing stone, but their second-ary position is evident. The keep form owes nothing to the Greeks.

From Saone southwards to Tortosa (Tartus) there are no early Latin fortresses: and Tortosa itself is one-half Byzantine, and the other half destroyed. Rey's plans (Fig. 29) give far more than can be seen today on the spot, for Tartus is flourishing, and heavy walls and rock-ditches are inconveniences in a growing seaport. The great square tower, which he marks within the innermost line of defence, is quite impossible to get at today, thanks to the existence of the harem of the governor[32] above it: what one can see from the streets suggests rather only two towers on the sea front.[33] Of course forty years ago there may have been more to find, but Rey's facts are sometimes persuaded to keep pace with his imagination, when the results are disastrous. The north gate of the city, which he describes, is standing however just as he left it—one of the most interesting gateways in the Latin East.[34] It is thirteenth century, but there can be no question as to the large amount of Byzantine remains in the two great walls of the town.[35]

[30] Most admirable latrines, with as usual a strong draught through them. [R] [Lawrence appears to have taken for a garderobe the small chamber from which the portcullis was operated above the entrance.]

[31] But it probably is. [X]

[32] Only a merchant I fancy after all. [R]

[33] [Compare the earlier draft of this section, below p. 126.]

[34] Rey, pp. 212–13, figs. 53–4[; Van Berchem and Fatio, *Voyage*, ii, pl. LXXIIIa. The gate-passage, about 8 ft. 8 ins. wide, contained a portcullis with a row of rectangular machicola-tions immediately in front of it.]

[35] Am going to stay in Tartus next year and do it well. A very interesting place. [X] [Lawrence overestimates the amount of Byzantine work—if any—surviving in the walls of Ṭarṭūs. Most of the features recorded by Rey (pp. 69–83, pls. VIII, XX) still exist, though obscured by modern buildings. Apart from the donjon, which is probably 12th century in origin, they seem to be mostly 13th century. More recent surveys of the castle and town walls

28. Saone. The ditch. [September 1909]

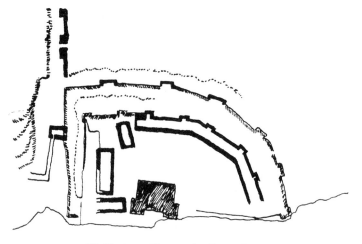

29. Tortosa. The castle (Rey, pl. xx)

Half a dozen miles southward of Tortosa lies Chastel Rouge (Kalaat Yahmur), a fortress without a history, and in its present condition an architectural enigma. It stands (Fig. 30) in a level plain, one of the very few flat places of northern Syria, without any natural advantages of position whatever: there have been ditches around it once, but they have been filled in completely,[36] so that probably they were never very large. The castle is composed of a square of walls, with rectangular angle-towers (Fig. 32). Two of the towers are corbelled out, but not pierced for *machicoulis*. So far the work could lay claim to Byzantine origin. Within this very small square of walls however is a Latin keep, of considerable dimensions, and most solidly built. It fills up nearly all the space within the containing wall[37] and therefore to this extent looks like an afterthought. Further the entrance to the keep is on the first floor, which proves that some building in front was understood when the keep was planned. There is now a kind of terrace on the western side of the keep, but not enough of it is visible to determine exactly whether it is the original arrangement or not (Fig. 32).

Otherwise the keep is not unlike Saone. The battlements are (or were) similar and the scheme of vaulting also corresponds. The stairs up are straight, but begin in the open air on the terrace. The plan (Fig. 31) will give all these features better than any description. The gate (Fig. 33) is interesting, from the close resemblance between it and the northern gate of Tortosa, in the placing of the *machicoulis*.[38] Those of the latter sort in this gate are comparable with those of Tortosa. Otherwise one can make nothing of it, and the foreworks beyond are amazing in their inefficiency: probably they are an addition of Arab times, and date from the period when the moat was filled in. If so the castle would be a compound of Crusader keep and gateway, of Byzantine outer walls, and Arab foreworks: and the unsuitability of the different parts, when arranged as they are, points rather at a composite effort. The keep is 58 feet long and 28 broad, and about 40 feet high.[39]

are found in C. Enlart, *Les Monuments des croisés dans le royaume de Jérusalem*, 2 vols. (Bibl. archéol. et hist., 7–8; Paris 1925–8), ii. 427–30, pls. 175–83; Müller-Wiener, *Castles*, pp. 50–1, pl. 33; Deschamps, *Châteaux*, iii. 287–92. Medieval Ṭarṭūs is the subject of a current research project undertaken by the German Institute in Damascus in conjunction with the Syrian Directorate of Antiquities and Museums: see M. Braune, 'Die mittelalterlichen Befestigungen der Stadt Tortosa/Ṭarṭūs: Vorbericht der Untersuchungen 1981–1982', *Damaszener Mitteilungen*, 2 (1984), 45–54.]

[36] [The uncorrected copy [X] reads, 'there may have been ditches around it once, but if so they have been filled in completely . . .']
[37] The rest is now filled with a warren of Arab huts, lived in by Arabs most squalid, and yet most suspicious. They refused to let me examine the keep at all closely.
[38] [See n. 34.] 'Machicoulis' are always used in two senses, the stone substitute for hoards, on a wall-top, and the square traps pierced in the floor of the room above an entry.
[39] [These dimensions, added after the Thesis had been typed, are different from those in the preliminary draft (54 × 73 ft.) (see below p. 128); nor do they correspond with those of

30. Chastel Rouge (Kalaʿat Yamur) [from north-east, August 1909]

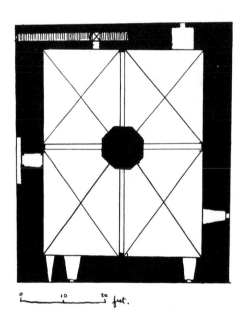

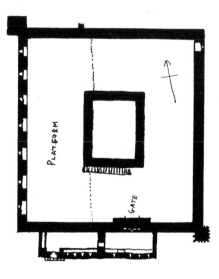

31. Keep of Chastel Rouge. First floor   32. Chastel Rouge (Kalaʿat Yahmour)

33. The gate of Kala'at Yahmur, from the foreworks. [Traced from a
photograph taken in August 1909]

33*a*. A little Metwali robber-hold in the Mseilha: fifteenth century probably, but no matter. [August 1909]

Inland from Kalaat Yahmur stands Chastel Blanc (Safita), the most elaborate of the keeps of northern Syria. The castle at Safita occupies the summit of a hill of considerable steepness, and the keep is on the highest point of all. Its dimensions are some 100 feet long by 60 broad, and it is still standing today, complete to its battlements, well over 100 feet above the poor houses of the village clustered round its foot (Fig. 36). The lower storey of the keep (Fig. 34) is used as the village church, as formerly it was the chapel of the castle. The door of entrance was blocked only by a single hinged gate, and appears a weak point: which is perhaps the reason why the door at the foot of the stair leading to the hall on the upper floor is also heavily barred and bolted. In the chapel there is one loophole high up above the altar, and two on each side of the nave. They are however so narrow, and at such a height above the ground as to be evidently unfitted for defence. Indeed as a rule loopholes are meant less for firing through than for admitting light: the recesses in which they are placed are seldom made high enough for a longbow, or broad enough for a crossbow. The archer would be forced to stand back within the tower; and to pass a yard-long arrow or broad quarrell through a slit 3 inches or less in width is a feat requiring some skill, even if the shot is to be straight ahead. Through such loopholes as those in the garden wall of New College, Oxford, an arc of fire[40] only 21 yards long at a range of 100 yards is the maximum. Walls generally speaking are always defended from the top.

The loops at Safita are evidently intended only to light the chapel, and they do it very badly at that. Those of the hall occupying the first floor (Fig. 45) are a little broader, but it is hardly possible from them to command the ground near the foot of the tower. The roof on the other hand is flat and unencumbered with fittings, and on the top of each merlon is a recessed socket, for the swinging bar of the shutter that closed the crenellations (Fig. 38):[41] this shows that some use was made of it in attacks, but even so the keep of Safita can never have been a very efficient stronghold. It would be crowded with a garrison of 200 men

Lawrence's own plan (Fig. 31), which, however, turns out to be considerably more accurate than the often reproduced sketch-plan published by P. Deschamps, *Les Châteaux des croisés en Terre Sainte*, i. *Le Crac des Chevaliers* (Bibl. archéol. et hist., 19; Paris, 1934), 56–7, fig. 11; cf. *Châteaux*, iii. 317–19, pl. LXXIXa; Müller-Wiener, *Castles*, p. 52, pls. 40–1. For a more recent survey, see D. Pringle, *The Red Tower (al-Burj al-Ahmar)* (British School of Archaeology in Jerusalem, Monograph Series, 1; London, 1986), 16–18, fig. 4, pls. I–III. The keep measures 53 by 46 ft. There is no evidence for Byzantine work in the castle, which seems to be largely 12th century with some Mamluk and possibly later additions.]

[40] An efficient arc. [X] [The sector of town wall at New College, Oxford, dates from the late 14th century: see RCHM, *Oxford*, 86, 159.]

[41] The battlements of Safita are the oldest extant. [X] [The claim is of course exaggerated, but Lawrence's enthusiasm is also communicated in a letter of Aug. 1909: 'a *Norman keep, with* ORIGINAL *battlements*: the like is not in Europe: such a find' (*Home Letters*, p. 104).]

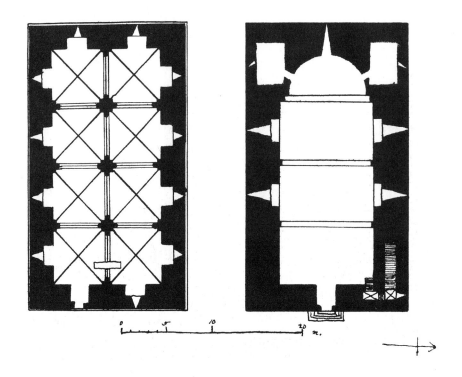

35. First floor   34. Ground floor

Safita, the keep
[Rey, fig. 26]

36. Safita from the east. [August 1909]

and the necessary stores, which must have included water, for the upper floor (evidently meant, by its barred door, as a last resort) has no means of access to the cistern in the foundations.

The keep can never have stood alone. There are remains of walls to the east, and in Rey's plan (Fig. 37) they are more comprehensible than today, since the governor of the district is grubbing them up to provide site and building material for his own residence. Below these yet further there is a ring-wall of polygonal shape with a very pronounced talus completely encircling the hill (Fig. 37). This might conceivably be Byzantine work, but some little carelessness in the construction allies it rather with the pseudo-Byzantine style adopted by the Templars in the late twelfth and early thirteenth centuries. Safita was one of the chief fiefs of Tripoli, and as such probably existed in Byzantine times, so that it is not impossible that the wall is earlier than the coming of the Crusaders. At any rate it is perfectly evident, from the difference in technique, that the builders of the keep had no hand in its construction. If it existed already, they took advantage of it: if not, they must have had something in its place. There would have been time between 1140 (supposing this to be the date of the keep) and 1220 for such an outer defence to have fallen into disrepair, or even to have become inadequate, for in Syria, with the constant fighting, development in military architecture came quickly.[42]

At Giblet (Gebal, Byblus, Jebeil) the square keep of exceptionally heavy masonry presents one or two features of late date. The gateway has a portcullis (Fig. 41), and the parapet is two storied, and of exceptional height, exactly like the inner curtain wall of a Byzantine fortress. Otherwise there is little to distinguish the keep from the others of its class, though the vault is managed entirely without pillars. The outer buildings (Fig. 39) are of Latin construction, and perhaps contemporary.[43]

These are the more important keeps yet standing in Syria. The remains of those at Botron (Batrum) and Caesarea are sufficient only to show that there had been a square keep there:[44] the keep at Beaufort (Kalaat esh Shukif) (Fig. 43) is of small size, and much damaged,[45] and

[42] [The verdict of P. Deschamps is that the keep was built by the Templars after 1171, while the 2 enceintes date from earlier in the 12th century. A 2-storey gate-tower on the east is assigned a 13th-century date: cf. *Châteaux*, iii. 249–58, pls. xxxi–xxxv. See also Rey, pp. 85–92, pl. ix; Enlart, *Monuments*, ii. 89–93; Müller-Wiener, *Castles*, pp. 51–2, pls. 36–9.]

[43] [See Van Berchem and Fatio, *Voyage*, pp. 105–8, pls. ii–iv; Müller-Wiener, *Castles*, pp. 64–5, pls. 85–7; Deschamps, *Châteaux*, iii. 203–15, pls. i–vii.]

[44] [On Batrūn (le Boutron), see E. G. Rey, *Les Colonies franques aux xiime et xiiime siècles* (Paris, 1883), 119, 363; Deschamps, *Châteaux*, iii. 9. The identification of a keep at Caesarea is also due to Rey (*Étude*, p. 222, pl. xxii), but little remains of it today: see Benvenisti, *Crusaders*, p. 143.]

[45] [Qal'at al-Shaqīf Arnūn: the keep, of 1139, was about 40 ft. square and 40 ft. high, with walls 9 ft. thick. The original entrance at ground-floor level was later replaced by one on the first floor. See Deschamps, *Châteaux*, ii. 204–6, 208; pls. lxxi–lxxii.]

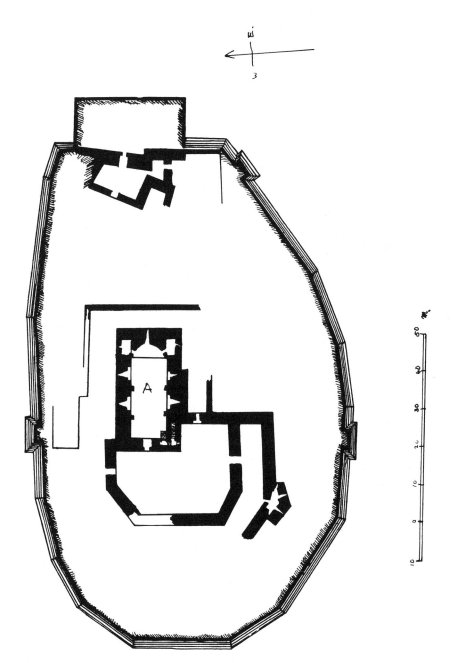

37. Safita (Chastel Blanc). A=keep. (Rey, pl. IX)

38. A shutter in embrasure   39. Giblet (Jebeil) (Rey, pl. xxı)

39a. Keep of Jebeil. X–Y: earth-bank: lined
stone revetment

40. Giblet. [August 1909]

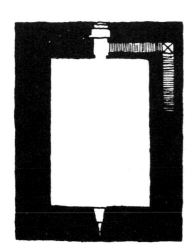

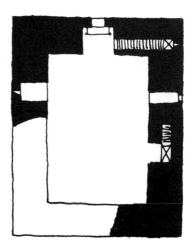

41. Ground floor   42. First floor

Giblet (Jebeil) (Rey, p. 119)

that at Banias in the great castle of Subeibeh (Fig. 44) on the hill above the springs of the Jordan has also suffered considerably. It had walls some 18 feet thick, and its dimensions are reasonably large (90 feet long by 83 feet broad).[46] The opportunities of the various Latin occupations of the place point to its being very early indeed in date.[47]

It will be evident from this summary sketch of the early Latin fortresses in Syria that the feudal nobility, who first were faced with the need for castle-building, allowed no interference with their plans of Greek architects. The shapeless, defenceless Norman keeps went up all over the country, with hardly an attempt at improvement. The parapets of keeps are sometimes a little elaborated, though at Safita the old simple fashion of Europe is maintained unaltered, and this is the largest, and on the whole the finest keep in Syria. One cannot as a rule say what the outworks of these keeps were like. It was not possible in the hills of central Syria to dig out a ditch, and line the mound inside with a palisade: for one thing there was no wood, and for another no earth. The chipping out of a moat like that of Saone in a basalt cliff demanded careful consideration and unlimited leisure, and the lack of this time and means persuaded the Latins, as might have been expected, to take over and utilise so far as possible existing Greek fortresses. In course of time, they put up imitations of Byzantine enceintes simply through force of custom: but it is evident that there was no rapid, complete abandonment of Western principles in fortification. The Normans and Provençals thought their keeps better than anything the Greeks could offer.

The general agreement seems to be that the square keeps in the East were built before the middle of the twelfth century. This date would suit fairly well with most of them. In Europe of course the form

[46] Very good. Wants more work. [X] [Lawrence visited Qal'at Ṣubayba in July 1909: 'on top of a hill about 700 ft. high . . . above the town [Banyas] is a much finer fortress. The view is extraordinary . . . Still it is a castle 500 yards long, on a spur of Hermon, and has got in one place rudimentary *machicoulis* like those of Château Gaillard, so I was very satisfied with it. I got all over the place, and at last set fire to the brushwood in the inner court which burnt all the morning. Still in the evening I profited, by seeing the building as a whole, as no other person can have done for 20 years—it was simply choked with rubbish' (*Letters*, p. 71; *Home Letters*, pp. 95–6).]

[47] Time of Reynier Brus, before 1135. [X] Before Reynier Brus in 1135. [R] [Renier Brus received Banyas from King Baldwin II in 1128, lost it to the Muslims in 1132, and regained it in 1139. Frankish occupation lasted until 1164. Ibn Shaddād al-Ḥālawī places the foundation of Qal'at Ṣubayba by the Franks in the period 524–7 H. (Dec. 1129–Oct. 1133); and it seems likely, as Lawrence's additional notes also imply, that this early work should be sought in the area of the keep or citadel. This is a rectangular enclosure, some 200 × 150 ft., flanked by 6 rectangular towers. Inside are remains of an earlier tower-keep, which has yet to be cleared and adequately surveyed. Lawrence's plan, copied from that made by C. H. C. Pirie-Gordon in 1908 (Fig. 44), is far from accurate. See Deschamps, *Châteaux*, ii. 173–4, pls. xl–xlii, l, plans; A. Graboïs, 'La Cité de Baniyas et le château de Subeibeh pendant les croisades', *Cahiers de civilization médiévale*, 13 (1970), 43–62 (pp. 58–61, fig. 5). See also n. 68.]

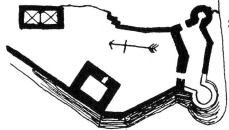

43. Beaufort (Kalaat esh Shu'kif) (Rey, pl. XIII)

44. Banias (Subeibeh). The keep

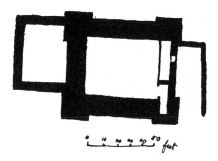

43a. [Beaufort Castle, July 1909]

45. Safita. The upper floor. From the north-east corner. The stairs from the
church come up behind the pillar in the midst, and at the back rise the stairs
to the roof

continued in favour a few years longer (though probably the number new-built decreased rapidly each year after 1150), but in Syria only Safita and Giblet appear to be as late: Giblet from its portcullis, and Safita from the piers and arches of its upper hall (Fig. 45).[48] On the other hand there is at present very little known about the development of Latin architecture in the East, so that a building which in France would be thirteenth century, might be twelfth century in Syria: and the mouldings and forms of the work at Safita are so plain that deductions from them are exceedingly delicate. One thing is quite certain: the keep form of castle in Syria is the earliest.

To determine what followed these keeps in the favour of the Latins is a more complicated question. Of course there were multitudes of castles built between 1150 and 1200, but for one thing, these buildings are hard to get at and, for another, when reached they have invariably been rebuilt. The only exception to the rule is the little hold of the Chastellet or Castle Jacob (Bet el Ahsan or Kalaat Gisr Benat Yakub) two miles down the river Jordan below Lake Huleh. It was begun, finished, taken, and destroyed in 1178–9, but unfortunately Saladin's troops razed it so completely that it is almost uninteresting. What does remain fairly visible is mainly a towerless curtain wall of no great strength.[49] Saladin probably found it much the same, since it was built in five months, and mined in five days. There was never more than a single line of wall.[50]

The group of castles around Tiberias, and as far to the north as Beaufort, are all known to have been built in the latter half of the twelfth century. They are Safed, built in 1140, rebuilt c.1180, and again c.1266, a Templar castle; Hunin destroyed in 1167 and 1187 and rebuilt by Baibars, with Safed, in 1267; Toron (Tibnin) built before 1105 in the first instance, and again after 1150, after 1218, and at some date in the sixteenth or seventeenth centuries; Belvoir (Kaukab el Hawa) built about 1180, taken from the Latins in 1188 and rebuilt since; and Tiberias, restored and rebuilt most frequently, till the last century. The great castle at Banias (Subeibeh) stands on a different plane from the others and must be considered with the rest of the Hospitaller fortresses. Of the others Toron and Tiberias are both hopeless, since the present buildings are not even on the foundations of the old.[51]

[48] [Jubayl is probably earlier and Ṣāfīthā post-1171: see nn. 42, 43.]

[49] But they scarped the rock. [X] [The existence of curtain-towers and the ruins possibly of a keep at the northern end of the site are recorded, however, by V. Guérin, *Description géographique, historique et archéologique de la Palestine*, iii. *Galilée*, 2 vols. (Paris, 1880), i. 341. Traces of large bossed masonry may also still be seen.]

[50] [The castle is described in Abū Shāmā, *Le Livre des deux jardins, Recueil des historiens des croisades: Historiens orientaux* (Paris, 1898), iv. 203–9.]

[51] [Of Tibnīn Lawrence wrote in Aug. 1909, 'the castle might have been worse' (*Letters,*

The first salient feature of all the other fortresses, however, is their very close imitation of Byzantine models. Hunin (Fig. 46) and Belvoir (Fig. 47) with their broad, though shallow rock-moats, and their very flat rectangular curtain-towers, are quite unlike any work of the period in Europe. They are rather forerunners of the later buildings of the Order of the Temple. It is quite reasonable to suppose that all the original masonry has been replaced at one time or other: but things that can never be destroyed are the rock-moats, and only the little depth that these possess proves them to have been by other hands than the Greeks.[52] The plans show practically everything that can be traced above ground, but this, though Greek in style, is all the more likely to be the work of Baibars.[53]

Safed is much more interesting. The castle hill (Fig. 48) is reasonably well covered with soil, and in consequence the ditches around the castle present a more European form. Earthquakes, and the expansion of the Jewish quarter between them, account for the disappearance of every stone in the building: fortunately a huge vaulted store-pit beneath the inner ward remains to prove the date of the place.[54] The entrance along the earthworks is very interesting: probably it crossed the moat by some sort of bridge resting on a tower that capped the

p. 72; *Home Letters*, p. 97). It is described by C. R. Conder and H. H. Kitchener, *The Survey of Western Palestine: Memoirs*, 3 vols. (London, 1881–3), i. 133–5, 207–8; cf. Deschamps, *Châteaux*, ii. 117–18, pls. xxxii–xxxiii. The existing castle and town walls of Tiberias appear to be entirely post-Crusader: see N. Feig. 'Tiberias', *Excavations and Survey in Israel*, 1 (1982), 10.]

[52] At Belvoir, Rey [*Colonies franques*, p. 437] declares that there are traces of a square keep inside the ditch and wall, and this of course if true would be somewhat puzzling: neither Mr Pirie-Gordon nor myself however could find the slightest trace of its existence. Rey was probably deceived by the wall of some Arab house. [Lawrence wrote in Aug. 1909 of Ḥūnīn, 'The castle there was trifling in strength, but as for fleas!' (*Letters*, p. 72; *Home Letters*, p. 97).]

[53] [The first Frankish castle at Ḥūnīn, built sometime after 1105, was destroyed by Nūr al-Dīn in 1167, rebuilt by Humphrey II of Toron in 1178, taken by Saladin in 1187, and demolished again by al-Muʿazzam ʿĪsā in 1222; it was refortified by the Franks in 1240 and taken and rebuilt by Sultan Baybars in 1266. As Lawrence indicates, very little masonry of the Frankish period remains (the large gate-tower is entirely Ottoman); but the rectangular plan and rock-cut moat seem very likely to represent the outline of the late 12th-century castle. See Conder and Kitchener, *Survey of Western Palestine*, i. 123–5; Deschamps, *Châteaux*, ii. 17, 128, 130, 134, pl. xxxiv; Benvenisti, *Crusaders*, pp. 300–3. Much more is now known of Belvoir (Kawkab al-Hawā), thanks to the clearances conducted in the 1960s. The outer rectilinear enceinte shown on Pirie-Gordon's plan (Fig. 47) and its outer rock-cut moat enclosed not a donjon (see previous note) but a rectangular inner ward with towers at its corners and a fifth containing a bent entrance on the west. The castle was constructed by the Hospitallers between 1168 and 1187. Some Ayyubid restoration work may have been done between 1189 and 1219, and possibly some Hospitaller reconstruction between 1241 and 1263 (after the dismantling of the castle by al-Muʿazzam ʿĪsā in 1228); but there is no evidence that Baybars contributed anything to it. See M. Ben-Dov, 'Belvoir', in *Encyclopedia of Archaeological Excavations in the Holy Land*, ed. M. Avi-Yonah (Oxford, 1975), i. 179–84; Benvenisti, *Crusaders*, pp. 294–300; J. Prawer, *The Latin Kingdom of Jerusalem* (London, 1972), 300–7; Smail, *The Crusaders in Syria and the Holy Land*, pp. 87, 100–2, pl. 9.]

[54] *c*.1150. [R] [But see n. 56.]

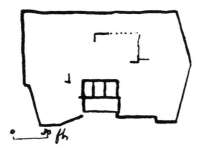

46. The castle at Hunin

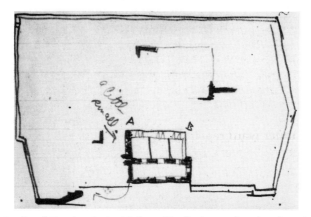

46a. Castle of Hunin. A–B, 30 feet. [Preliminary drawing for Fig. 46]

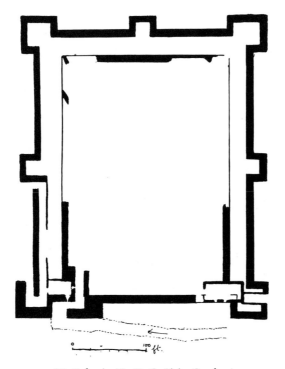

47. Belvoir (C. H. C. Pirie-Gordon)

mound E. There were evidently other towers (rectangular),[55] restored conjecturally in outline, along the top of the mounds. If complete, Safed would have been one of the most valuable fortresses in Syria. It belonged to the Templars.[56]

Professor Oman quotes from William of Tyre the description of Darum in the southern coasts of Palestine, built by Amaury about 1160 on a purely Byzantine plan, like the outworks of Giblet. There is now nothing whatever left of it: 'sed [tamen] absque vallo erat et sine antemurali', and its weakness is sufficiently shown by Richard's storm of it in four days in 1192.[57] Rey attempts to associate it with Blanche-Garde and Ibelin, but excavations at Blanche-Garde some years ago proved his plan wholly imaginary.[58]

These few not very important castles have been dwelt on because they fill the gap between the square keeps and the fully developed fortresses of the two great Orders: and in filling it they show incidentally how entirely the East had lost touch with the West in military engineering. There is no castle in France with the least resemblance to any one of them, but there are numbers most clearly related all along the Byzantine frontiers. If these castles were the last produced by the Latins, the classical view as to their entire absorption in Byzantine ideas

[55] Pirie-Gordon says circular: and sticks to it. Will go once more and dig it up. [X] [Both were probably right, since the towers of the inner ward seem to have been rectangular, those of the outer rounded and the Mamluk keep cylindrical: see D. Pringle, 'Reconstructing the Castle of Safad', *Palestine Exploration Quarterly*, 117 (1985), 139–49. The rectangular tower to which Lawrence refers in his preliminary draft (see below p. 128) may have been a tower of the inner ward or possibly part of the Ottoman work enclosing the Mamluk keep; but it should not be confused with the keep itself.]

[56] [Lawrence spent several days in Ṣafad in July 1909. 'From the castle (good) two-thirds of the Holy Land is visible. I had great sport in an underground passage that is being excavated: xiii century work I think' (*Letters*, p. 73; *Home Letters*, p. 98). The castle was founded in 1102, rebuilt in 1140, passed to the Templars in 1168, fell to Saladin in 1187 and was destroyed by al-Muʿaẓẓam ʿĪsā in 1218–19. Between 1240 and 1260 the Templars made it one of their most formidable strongholds in Palestine; and after the castle fell to him in 1266, Baybars and his successor Qalāwun worked at strengthening it still further. The fragmentary remains therefore belong to various periods and only excavation can resolve the difficulties in interpreting them. For 2 recent attempts, see R. B. C. Huygens, *De Constructione Castri Saphet: Construction et fonctions d'un château fort franc en Terre Sainte* (Amsterdam, Oxford, New York, 1981); and Pringle, article cited in previous note.]

[57] William of Tyre, xx. 19 [*Recueil des historiens des croisades: Historiens occidentaux* (Paris, 1844), i. 975 (trans. E. A. Babcock and A. C. Krey, *A History of Deeds Done Beyond the Seas*, 2 vols. (New York, 1943), ii. 372–3). King Amaury reigned from 1162 to 1173. Darom had been enlarged by 1192, for while Amaury's castle had only 4 towers, the castle stormed by King Richard I had 17 (Deschamps, *Châteaux*, ii. 14–15). The site has still to be found beneath the dunes.]

[58] Rey, pp. 123–5, fig. 39[; cf. F. J. Bliss and R. A. S. Macalister, *Excavations in Palestine During the Years 1898–1900* (Palestine Exploration Fund; London, 1902), 28–43, pl. 7, figs. 10–12. Little remains of the castle of Ibelin (Yibnā) save for the foundations of what may perhaps have been a tower-keep. Both Blanchegarde and Ibelin, however, are described by William of Tyre as rectangular castles with towers at the corners, built in 1141 and 1142 respectively (xv. 24–5; *RHC Occ.* i. 696–8 (trans. ii. 130–2)).]

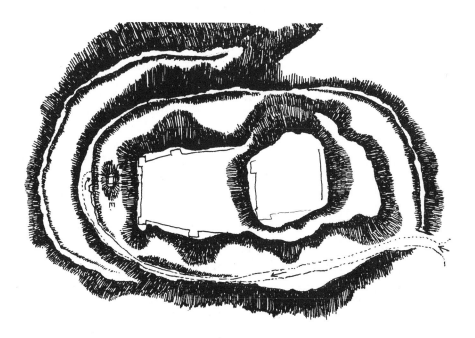

48. Safed. [August 1909]

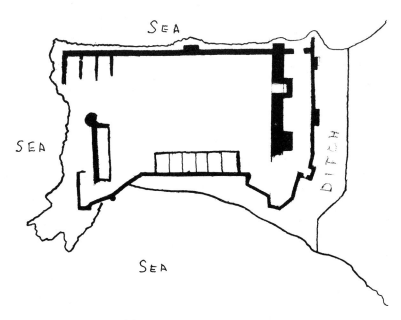

49. Château Pèlerin (Athlit) (Rey, pl. x)

would be more than justified: though it would be worth noting that the absorption took over half a century to bear fruit.

The establishment of the Military Orders in all the frontier fortresses of the Latin Kingdom meant a new era in Syrian castle-building. It is only too evident in studying the works of the private holders of fiefs, even of important ones, that a lack of material resources, of money and labour, hampered their efforts most cruelly. Another obstacle to elaborate defence-works was the insecurity of tenure. Families were continually dying out, with the abnormal death rate of Palestine through disease and accident, and besides this, fiefs were continually transferred. The Military Orders were ideally fitted for such conditions. The members were celibate, and so, easy to control, and without private interests: they had no heirs to search after, and no domain to preserve intact. Then the Orders were everlasting, with an inexhaustible supply of the finest chivalry in Europe to draw upon in case of need. The military ability of the commanders, and of the simple members of either Order is again and again brought out in striking contrast with the inefficiency of the laymen of the Kingdom. There was a tradition, after a little, among the knights, of the conduct of warfare against the Infidel, and each newcomer of repute vivified this tradition with the fruits of his own talent and experience. Most important of all, perhaps, the Orders were very rich, not only in the precarious possession of one-half the most fertile land of Palestine, but in property in Europe; property which would maintain the prosperity of the Order when a sudden raid had set the ordinary nobles of Syria face to face with financial ruin.

Perhaps all these many and varied advantages of the Orders were most felt where they have survived most clearly, in their military architecture. The Orders held practically every fortress of importance in northern Syria, and they added to or rebuilt nine-tenths of those they occupied. This building was done by Templars and Hospitallers in the same period, the last twenty years of the twelfth century, and the first half of the thirteenth century, but the bitter rivalry and jealousy between them led them to adopt different styles for their buildings. The Templars, always suspected of a leaning towards mysterious Eastern arts and heresies, took up the mantle of Justinian, as represented by the degenerate fortresses in northern Syria, and amplified it, in making it more simple. The Hospitallers, in harmony with their more conservative tradition, drew their inspiration from the flourishing school of military architects in contemporary France, and so the inborn antipathy of East and West, which more than any one thing has been the primary cause of all Crusades was demonstrated in the fortresses of the Latin East. The two schools of builders had entirely different ideas and principles, and the two classes of buildings are entirely distinct, without a link or compromise between them. It will be easier to consider first the castles of the Templars, as the smaller class, and the one least fruitful of results.

The characteristics of the Templar style will be grasped at once if a plan (Fig. 49) of Château Pèlerin (Athlit) their chief stronghold be considered. They held possession there of a narrow promontory of rock and sand, eminently defensible according to medieval ways. Yet here the Templars, working in 1218, threw aside all the carefully arranged schemes of flanking fire, all the covering works, all the lines of multiple defence which were being thought out meanwhile in Europe. At Athlit they relied on the one line of defence—an enormously thick wall, of colossal blocks of stone, with two scarcely projecting rectangular towers upon it. These were the keeps, the master towers of the fortress, and instead of being cunningly arranged where they would be least accessible they are placed across the danger line, to bear the full brunt of the attack. One would expect them to be unusually massive, but they are, in true Byzantine style, of thin walls, compared with their curtain,[59] and the hoard, which was just then being generally adopted, is not made use of to repair the weakness. The projection of these towers is very slight, insufficient to rake an enemy busied on the face of the curtain, and the little προτείχισμα in front is not of a force to be held alone. The strength of Athlit was brute strength, depending on the defenceless solidity of the inner wall, its impassable height, and the obstacle to mining of a deep sea-level ditch in the sand and rock before the towers. The design is simply unintelligent, a reworking of the old ideas of Procopius, only half understood. Justinian, except in rare exceptions, had not intended his fortresses to stand alone, as the last refuge in a conquered country: they were temporary defences to assist the unrivalled Greek field army. Given unlimited time and labour, anyone can make a ditch so deep and a wall so high of stones so heavy as to be impregnable: but such a place is as much a prison for its defenders as a refuge: in fact a stupidity. Such is Athlit.[60]

Of the other still preserved Templar fortresses, that at Areymeh (Fig. 50) is a little better. The Templars there inherited a Byzantine site, and merely rebuilt the inner ward. It stood on a hill so precipitous that

[59] One in Fig. 49 is shown solid: this is because only the foundations are left. [X] [The front walls of the towers are nevertheless 24 ft. thick and the curtain about 40 ft.]

[60] [The sophistication of the Templars' triple line of defences at ʿAtlīt, which did indeed make use of hoards and machicolations to provide flanking and covering fire, had to wait until the survey and excavations directed by C. N. Johns between 1931 and 1936 to be fully appreciated. See C. N. Johns, 'Excavations at Pilgrims' Castle (ʿAtlit)', *Quarterly of the Department of Antiquities of Palestine*, 1 (1932), 111–29; 3 (1934), 145–64, 173; 4 (1935), 122–37; 5 (1936), 31–60; id., *Guide to ʿAtlit: The Crusader Castle, Town and Surroundings* (Jerusalem, 1947); id., 'ʿAtlit', *Encyclopedia of Archaeological Excavations in the Holy Land*, ed. M. Avi-Yonah (Oxford, 1975), i. 130–40.]

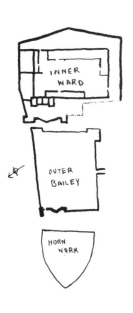

50. Areimeh, much destroyed

50a. Areymeh. [A tower of the inner ward, August 1909]

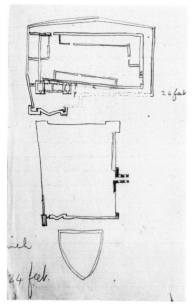

50b. Areimeh. x–y = 24 feet. [preliminary drawing for Fig. 50]

51. Safita. Girdle-wall. [August 1909]

attack could only be delivered on the horn-work to the west: and after that had been carried there was an outer ward, cut off from the inner ward by a ditch. The plan explains easily the arrangement of the place. One might wish only for some curtain-towers in an attack. Parts of the wall are a little bare against unexpected escalade, above all as they are mostly of very poor height. The horn-work too is not intended seriously.[61]

The Templars, in their very early days, built Safed, where nothing is left but an admirable series of ditches; probably also the polygonal wall (Fig. 37) that girdles the hill of Safita is due to them. It is a low wall today (Fig. 51), but may have been once comparatively lofty, and it has a very great talus. Its numerous dead angles, undefended, are however the reverse of attractive. Tortosa was another great fief of the Templars: they seem to have occupied it in 1180, or the year after, when the great square tower, described in the siege of the place after Hattin, was already standing. For the rest, Tortosa is defended (Fig. 29) by semi-circular ditches and slight curtain-towers, apparently of the Byzantine age, for it was a fortress when the Latins came, and the present arrangement of ditches is almost the only possible one. The walls are little credit to their designer.[62]

It is more pleasant to turn to the buildings of the knights of the Hospital, finding the beginnings of their work in parts of their castle of Banias, and tracing it down through Crac and Markab. From the beginning it is a style absolutely different from the pseudo-Byzantine in fashion in Syria when it was introduced, and it left no legacy behind it in the Arab fortifications of a later date, except conceivably in the box-*machicoulis* so common in Arab work of the late thirteenth century.

Banias, which is the earliest Hospitaller castle that can be certainly dated, was destroyed by El Mu'azzam in 1218, after the Hospitallers had lost it finally in 1164. It was a much-disputed place, continually besieged by one party or other, on account of its very formidable position over the great Damascus road, and the springs of Jordan. The Franks first won it in 1129, and they held it for three years. During this time unquestionably they erected the large square keep, with pilasters at the angles, the only instance of such in Syria.[63] The Arabs only held it one year, and then the Latins recovered it, and handed it over to the

---

[61] [On 'Arayma, see Müller-Wiener, *Castles*, p. 53, pls. 42–3; Deschamps, *Châteaux*, iii. 313–16.]

[62] [Ṣāfīthā: see n. 42; Tarṭūs: see n. 35; Ṣafad: see n. 56.]

[63] See earlier [p. 62, and n. 47.] [Lawrence is writing here of Qal'at Ṣubayba, not of Banias itself. Of the latter he wrote in July 1909, 'The town was formerly fortified, but mostly Roman and Arab work I fancy' (*Letters*, pp. 70–1; *Home Letters*, p. 95). On these defences, which are mostly Ayyubid and Mamluk, see Benvenisti, *Crusaders*, pp. 152–4; Graboïs, *Cah. de civ. méd.*, 13 (1970), 48–55.]

Hospitallers. The latter had thus twenty years of unbroken, though not undisturbed, occupancy. The plan (Figs. 52, 53) makes it evident that the large round tower on the south-east of the keep, and the gallery beside it, are later additions. The style in which these and the round towers next to it, going westwards, are built is undoubtedly Christian, and equally certainly European in origin. The Arabs never employed round towers themselves, and there are other Christian features in the building that make a Mohammedan origin impossible.[64] It is worth noting also that the Hospitaller rebuilding has remained quite incomplete. The square keep and Byzantine-like forework and ditch are of one period homogeneously enough, and the four circular or semicircular towers: the rest is Arab, and of comparatively later date. The rebuilding of the castle must have been proceeding when the Arabs recovered it, and this, if correct, would put the part remaining about the year 1160. On one of the towers D (Fig. 53) are some *machicoulis* (Fig. 54) and they were evidently in the original plan:[65] there are no signs of rebuilding. Their pattern is not quite an ordinary one, since there were only a few, spaced with wide intervals around the top of a very large tower. At the same time they were unquestionably *machicoulis*, and not at all rudimentary in design. The common view of course is that *machicoulis* were invented in Syria owing to the lack of wood for hoarding; but on the other hand they are comparatively rare in the East and the Templars, who represent the native style of building, never adopted them at all.[66] The designs of the buildings of the Hospital were on the French model, and these *machicoulis* have a very distinct Provençal or north Italian feeling.[67] Banias was probably the first castle the Hospitallers put up in Syria, and yet their architect must have had frequent opportunity of building such *machicoulis*, since only by use could their distinctly decorative quality have been developed. Similar ones do not appear on any other Crusading castle in the country: and when the Arabs adopted *machicoulis* they were not of this shape.[68]

[64] The gate B is what C. H. C. Pirie-Gordon has named the 'Hospital' gate.

[65] The earliest *machicoulis*. [X] [See n. 68.]

[66] [Corbelled Templar *machicoulis* are recorded at ʿAtlīt. They evidently existed elsewhere but have simply not survived.]

[67] According to Mr [W. R.] Lethaby who saw the photograph. [In a letter of July 1909 Lawrence had compared them to Château Gaillard (see n. 46).]

[68] [On the contrary, these *machicoulis*—and all the rounded towers at Qalʿat Ṣubayba—may now be dated to the Ayyubid and Mamluk refortification from 1228 onwards. The extent of 12th-century Frankish work in the castle is still uncertain and may have concerned no more than the keep (see n. 47). It is grossly overestimated by P. Deschamps and his architect P. Coupel (*Châteaux*, ii. 145–74, plans); and one writer has even raised doubts as to whether any part of the castle can be attributed to the Franks (Benvenisti, *Crusaders*, pp. 147–9, 154–7). Be that as it may, it seems unlikely (*pace* Graboïs, *Cah. de civ. méd.*, 13 (1970), 46, 60–1) that any of it is the work of the Hospitallers. The Hospitallers received half of Banias from Renier Brus in 1157; but their convoy was ambushed and annihilated by Nūr al-Dīn before it ever reached

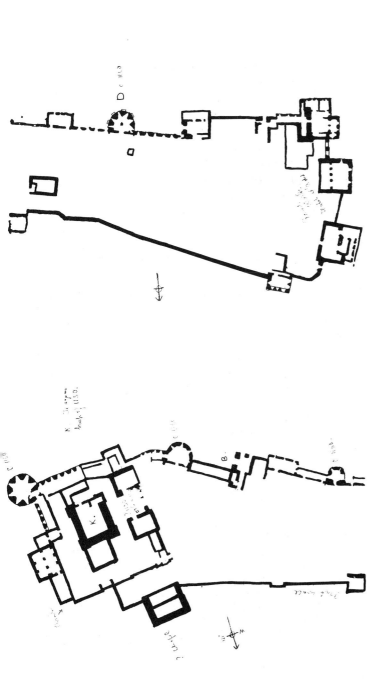

53. Banias (Subeibeh)
Western half
(C. H. C. Pirie-Gordon)
[Rounded Towers:] c. 1160
[Rectangular towers at west front:] Arab work. Malek el ʿAdel

52. Banias (Subeibeh)
Eastern half
(C. H. C. Pirie-Gordon)
K. The square keep of 1130
[Rectangular towers of inner ward:] ?1130 or Greek
[Rounded towers:] c. 1160

54. *Machicoulis* at Banias [Ṣubayba: July 1909]

55. Crac des Chevaliers from the west. Behold the camels in front [August 1909]

The other two great Hospitaller castles in the north, Crac des Chevaliers (Hosn el Akrad, Kalaat el Hosn) and Margat (Markab), both date mainly from the thirteenth century. At the same time Crac, as a finished example of the style of the Order, and perhaps the best preserved and most wholly admirable castle in the world, forms a fitting commentary on any account of the Crusading buildings of Syria.[69] It cannot compare for a moment with Coucy in France, or with Caerphilly[70] in its science of defence, but is more impressive than both since it is neither ruin nor showplace. A few years back it withstood a siege on the part of a neighbouring district with complete success, and were Baibars to reappear he would think it as formidable as of old.

There are signs of two or three periods of construction (Fig. 56). In the inner ward the wall from the tower of entry L to the tower with buttress-*machicoulis* P, including the chapel M, appears earlier than the rest of the inner ward. Of the outer ward the whole south front is Arab, and the western side so far as tower D. In the siege of the place in 1271 by Baibars the central tower A was entirely destroyed, and probably other parts of the outer line as well.[71] The lower part of the three half-round towers on this stretch of wall is old. The upper parts were rebuilt by Baibars, and Malek-es-Said Bereke-Khan.[72] Kelaoun[73] built the square tower A, and Baibars the gate-tower D¹. The rest of the outer ward is Hospitaller work, though later in date than the inner ward.

the town. There is no evidence that they ever garrisoned it or Ṣubayba before the town fell in 1164 (William of Tyre, xviii. 12; *RHC Occ.* i. 837 (trans. ii. 256–7)). The failure of the medieval sources to distinguish clearly between the castles of Banyas and Ṣubayba is one of the principal reasons for this confusion; only a major review of these sources and of the archaeological and epigraphic evidence seems likely to produce a satisfactory answer.]

[69] [Lawrence stayed 3 days in Crac in Aug. 1909, describing it as 'the finest castle in the world: certainly the most picturesque I have seen—quite marvellous' (*Letters*, pp. 76, 95; *Home Letters*, p. 104). At that time the castle was still occupied by a village, and much of the medieval masonry was obscured by house walls and other debris. Some fine photographs of its appearance a decade before his visit may be seen in Van Berchem and Fatio, *Voyage*, ii. pls. x–xxi; cf. i. 135–62; cf. G. Bell, *The Desert and the Sown* (London, 1907), 201–7. Clearance work in 1927–9 and the subsequent resiting of the village outside the castle (1932) have enabled a more thorough examination of the building's plan and phases of construction than was possible to Rey (pp. 39–67, pls. IV–VII). The fundamental description and analysis of the castle is now P. Deschamps, *Les Châteaux des croisés en Terre Sainte*, i. *Le Crac des Chevaliers*, text and album (Bibl. archéol. et hist., 19; Paris, 1934). Useful shorter accounts include the same author's *Le Krak des Chevaliers* (Gazette des Beaux Arts; Paris, 1929); Müller-Wiener, *Castles*, pp. 59–63, pls. 66–83; and A. Rihaoüi, *The Krak of the Knights: Touristic and Archaeological Guide* (Directorate General of Antiquities and Museums; Damascus, 1982).]

[70] [See C. N. Johns, *Caerphilly Castle, Mid Glamorgan* (HMSO; Cardiff, 1978).]

[71] [On the siege of 1271, see Deschamps, *Châteaux*, i. 132–6; D. J. Cathcart King, 'The Taking of Le Krak des Chevaliers in 1271', *Antiquity*, 23 (1949), 83–92.]

[72] [Al-Malik al-Saʿīd Muḥammad Baraka Khān, the son of Baybars, became joint Sultan in Aug. 1264 and abdicated in Aug. 1279.]

[73] [Al-Manṣūr Qalāwun, Mamluk Sultan, 1279–90.]

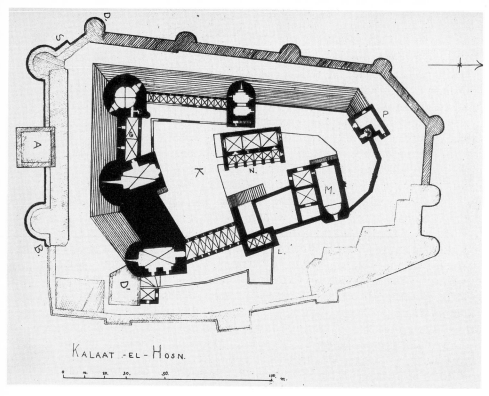

KALAAT -EL- HOSN.

56. [Crac des Chevaliers, after Rey]

57. Crac from the east. [August 1909]

The castle is entered (Fig. 58a) at D by means of a plain gate, and then a vaulted passage, almost dark, as far as the tower J, where is a large trap-*machicoulis* in the roof and other defences.[74] To reach the inner ward one must go further, to the tower L, also in a dark vaulted passage, ascending steeply. To surprise this entry would therefore be extremely difficult.[75] There are only three or four loopholes to give light and air, and the change from the glare of the sun without is most confusing. The upper gate has a portcullis, but generally it is very like the 'Hospital' gate at Banias: through it one enters a small courtyard, opposite the great hall of the castle N.[76] Another flight of stairs gives on the upper court K still unencumbered with houses (Fig. 56). From it lead up more steps to G (Fig. 59), the platform uniting the three great towers that together form the donjon. They overtop by many feet any other tower in the fortress, and are magnificently built of huge blocks of stone. The governor of the province now inhabits them, and his harem and his divan, and his own private rooms rather obscure the arrangement of the eastern half of the platform. The western half of it, G^1, was however vaulted (Fig. 59).[77]

From these towers the great wall, known to the Arab historians as 'the mountain', slopes outwards and downwards for more than 80 feet to the thick greenish mud and water of the moat (Fig. 60). Below tower E it runs out in an enormous spur and then at right angles it turns along the whole western front until it is lost in the rectangular tower P (Fig. 61).

The reason for making the wall with so great a batter and such thickness[78]—nearly 80 feet—is a little hard to find. Against an earthquake it would be useful perhaps, though no part of Crac has been damaged by one: the castle stands on rock, so mining was not greatly to be feared: and half the thickness would have been secure against any ram that ever was imagined. It had however one advantage against ordinary attacks in the absence of *machicoulis*: assailants could never get underneath the fire of the defenders on the fighting platform: and this after all may have been the real purpose of the construction. On the other hand it had the drawback of making easy escalade.[79]

[74] [For a more accurate survey of the entrance to the castle, see now Deschamps, *Châteaux*, i, plan 1.]

[75] It is not easy even today to stumble up the uneven steps, in a litter of pariah dogs and goats. (Also cows and fleas, etc. [R].)

[76] [Now identified as the Chapter House.]

[77] [This area, particularly the richly decorated south-western tower, was probably reserved for the Hospitaller commander of the castle: see Deschamps, *Châteaux*, i. 209–11, 289–90, pls. XLI, XCV, plan 6K; Enlart, *Monuments*, ii. 98–9. The governor's apartments are described by G. Bell (see n. 69) and also by Sir Harry Luke, who, together with C. H. C. Pirie-Gordon, had been a guest there in 1908, a year before Lawrence: see *Cities and Men*, 3 vols. (London, 1953), i. 149–50.]

[78] If turned over it would be the same height. [R]

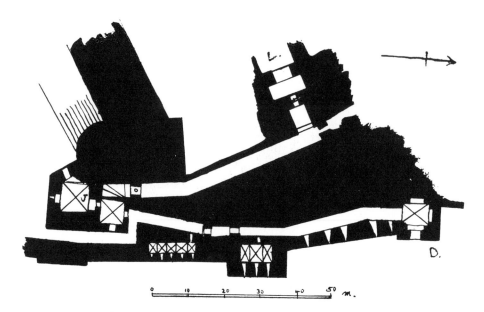

58*a*. Kalaat-el-Hosn (Crac des Chevaliers). The entry (Rey, p. 47)

59. The platform above the talus on south, Crac des Chevaliers [August 1909]

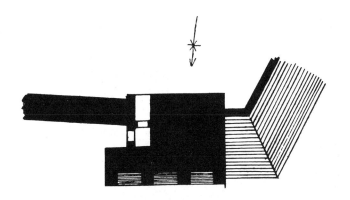

58*b*. Kalaat-el-Hosn (Crac des Chevaliers). Tower P (Rey, p. 57)

58*c*. Kalaat-el-Hosn (Crac des Chevaliers). Tower P (Rey, p. 57)

60. The great talus: Crac des Chevaliers: looking east. [August 1909]

In the matter of *machicoulis* Crac is most comprehensive. The ordinary pattern, as used generally from the thirteenth century in France, is employed in the outer line of wall from tower S to tower B (Fig. 56), and elsewhere on the outer line. Then from tower S northwards the wall contains near the top a vaulted gallery, with openings at intervals into small chambers, corbelled out from the face of the wall (Figs. 62, 63). They resemble the latrines common in France in appearance, but are defensive in intention. Each is of a size fitted for one man, or two at a pinch, but freedom of movement would be very severely hampered when working in a room only 16 inches wide. No kind of bow could be used. The openings would be available only for dropping stones. If a gallery in a wall is to have *machicoulis* this is of course the only possible pattern: but the whole thing is not very effective to Western minds, accustomed to the unbroken ring of corbelling along the top of the wall. In the East, however, the tops of towers could not be roofed in,[80] and so the covered *machicoulis* were the better form. The Arabs adopted them wholeheartedly in Aleppo, and Damascus and elsewhere, and presumably the credit of their introduction lies with the Hospitallers of Crac des Chevaliers. At least they have not been found in earlier buildings of the Latins.[81]

A third and still more interesting form of *machicoulis* exists on the tower (P) of the inner ward (Fig. 65). It is composed of buttresses applied to the face of the tower, and arched over at a height of some 30 feet. The front wall of the tower is then carried on them, with a peculiar double system of relieving-arches relieving nothing. On top of all seem to have been *machicoulis* of the corbelled pattern (Fig. 58*b* and *c*). The mason-work of the tower is quite like that of the great sloping wall, but nothing parallel with these *machicoulis* is known in Syria:[82] and in

[79] I was able, barefoot, to climb up more than half-way (about 40 ft. [X]) though with some little difficulty (a gentle way of putting it: and I had to come down [R]). [The batter, or glacis, is dated by Deschamps to the second Frankish period of construction, towards the end of the 12th century and the early part of the 13th, following an earthquake which did serious damage to the castle in 1170 (*Châteaux*, i. 121, 281). The outer enceinte was also added in this phase (pp. 279–83).]

[80] No timber or slates. [X]

[81] [More likely, as Deschamps argues (pp. 262–6), it was the Franks who borrowed this type of machicolation from the Arabs. The machicolated galleries on the south front seem to be entirely Mamluk. The box-*machicoulis* on the western side are Frankish of the 13th century and may, as Deschamps suggests (p. 262), represent a translation into stone of the wooden hoards employed in the West; but stone-built *machicoulis* of this type are also found in Muslim architecture in the 8th century (see Creswell, *Proc. Brit. Acad.*, 38 (1952), 91, pl. 1).]

[82] Miss Bell has lately found some rather like at Uk-heidar in the desert of Baghdad in a Sassanian building. [R] One was found lately by Miss G. Bell in a 7th-century palace near Baghdad. [X] [The palace, Ukhaiḍir, is 8th century and Muslim: see G. L. Bell, *Palace and Mosque at Ukhaiḍir: A Study in Early Mohammadan Architecture* (Oxford, 1914), 7, 107, 121, pls. 5.3, 7.2, 8.1, 9.2.]

61. The west face of the inner ward: Crac des Chevaliers

63. Box—*machicoulis*, Crac des Chevaliers. [August 1909]

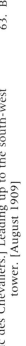

62. [Crac des Chevaliers.] Leading up to the south-west tower. [August 1909]

65. The tower [P] of the inner ward, Crac des Chevaliers. [August 1909]

64. [Crac des Chevaliers. Tower P from north-east. Cf. Fig. 58b. August 1909]

67. Markab, looking east on the southern [south-east] face. A sort of beehive underneath is a sheikh's tomb: the steps of the entry are about 18 ins. high. [August 1909]

66. Markab, looking south[-west], along the south-east wall. August 1909

Europe very few exist of the same pattern (p. 109). They ought to be immensely superior to the ordinary tiny *machicoulis*, for the size would allow if necessary of a whole beam of wood being thrown down. It was perhaps less stable—the destruction of the lower part of the buttresses might cause the whole to collapse—but whether for this or other reason it never found favour. In this particular instance a postern is concealed behind it.[83]

Markab is much more French even than Crac. The outer work on the east was rebuilt by Kelaoun,[84] but all the rest might be a part of unrestored Carcassonne. The narrow lists between the walls (Figs. 66, 67), the chapel, and the great round tower (Fig. 66) are simply typical of the best period of French architecture. To describe it is not necessary, for Rey's plans and drawings[85] present a faithful picture of the main features of the castle: and he also emphasises the thirteenth-century character of the whole. It was only right that the creator of the 'last word in Syrian castle-building'[86] should return to the West, both for the general design of his fortress, and the particular architectural details; just as the first builders, the men of Saone or Safita had done. The only people wholly independent of Europe had been the Templars, and their style was practised only by themselves, and died with them. All the best of the Latin fortifications of the Middle Ages in the East was informed with the spirit of the architects of central and southern France.

VI. MILITARY ARCHITECTURE
IN EUROPE IN THE TWELFTH
CENTURY

PROGRESS in military architecture in Europe depended on the elimination, or at least the modification, of the square keep form, and accordingly from 1150 there are on record many and varied attempts to improve it, or to give it less prominence in the scheme of defence.[1] The

[83] [The tower is described by Deschamps, who dates the addition of the buttress-*machicoulis* to the late 12th century (after 1170) and the upper gallery-machicolation to the period 1255–65. See *Châteaux*, i. 185–7, 265, 284–5, fig. 35, pls. XLVII, XCIIb, XCIII.]

[84] [The castle fell to Sultan Qalāwun in 1285.]

[85] [Rey, pp. 39–67, figs. 9–19, pls. IV–VIII. See also Deschamps, *Châteaux*, iii. 259–84, plans, pls. XXXVI–LVI; Müller-Wiener, *Castles*, pp. 57–8, pls. 52–61, frontispiece; Ahmad Fāʾiz al-Hamsī, *Qalʿat al-Marqab* (Directorate of Antiquities and Museums; Damascus, 1982).]

[86] [C. H. C.] Pirie-Gordon. [R]

[1] [See P. Héliot, 'L'évolution du donjon dans le nord-ouest de la France et en Angleterre au XIIᵉ siècle', *Bulletin archéologique du Comité des Travaux historiques*, NS 5 (1969), 141–94.]

FRANCE
PLACES VISITED
1907. 1908.

Map II. France. Places visited by T. E. Lawrence, 1907, 1908

idea that a donjon or last resort of some kind was absolutely necessary persisted strongly through the greater part of the thirteenth century: probably only the introduction of cannon drove it from the field. Long before this time, however, the capture of the outer and inner wards of the castle had meant the surrender of the whole.

Modification of the Norman keep took two forms. In the one case a donjon was retained as the more important part of the castle, though its shape was not a perfect square with undefended angles: in another the great keep of passive resistance was swept away entirely, and a light shell keep, with or without a little tower inside it, took its place. In the second instance as time went on the size and importance of the shell increased: until it formed the inner ward of the castle, and the tower inside became the donjon. Château Gaillard is the finished result of this process. In other overgrown shell-keeps there is no trace of any tower within them at all: in fact the donjon idea has been eliminated. Château Gaillard did not attain to this pitch of perfection, though in practice its keep proved useless.

The modification of the first kind appears to have begun with some such process as that evident at Mitford near Morpeth, where one wall of the keep has been thrown forward in an obtuse angle.[2] At Chalusset near Limoges (Fig. 69) this spur is made practical use of, to provide extra thickness of wall on the most vulnerable face. The keep here is of course a very small one, but of quite a normal pattern for the centre of France.[3] Huge keeps like Falaise or Arques are not found south of the valley of the Loire:[4] they become narrow as at Luzech or tall and slight as at Marthon or St Yrieix.[5] This example at Chalusset is quite early, and is perhaps the forerunner of the towers with spurs on weak faces, a type which appears in the Lower Seine valley at La Roche Guyon (Fig. 70)[6] and Château Gaillard (Fig. 71) and elsewhere. They bear no relation to the half-hexagonal towers sometimes found in Byzantine fortresses, since there the projecting point is not exceptionally strong. There was no question of resisting with it blows of a battering ram.

[2] [See Renn, *Norman Castles*, pp. 245–7, fig. 47.]

[3] [Lawrence writes in a letter of Aug., 'Chalusset, a *most wonderful* thing of the xiii cent. [A] fine castle, with donjon of xii cent. *and a large beak on the front of it.* "Eureka" I've got it at last for the thesis: the transition from square keep form: really it is too great for words' (*Letters*, p. 61). The keep dates from 1132: see A. Demartial, 'Château de Chalucet', *Congrès archéol. de France*, 84 (1921), 261–8; Châtelain, *Donjons romans*, pp. 196–8, pls. xv, xliv.]

[4] [The keep at Falaise was built by Henry I, *c.* 1123: see M. Deshoulières, 'L'âge du donjon de Falaise', *Bulletin monumental*, 85 (1926), 204; R. E. Doranlo, 'Le château de Falaise', *Congrès archéol. de France*, III (1953), 181–200; Châtelain, *Donjons romans*, pp. 118–20, pls. iv, xxiii. For Arques, see p. 22 n. 44, and Châtelain, *Donjons romans*, pp. 113–14, pl. iv.]

[5] [See R. Fage, 'Saint-Yrieix', *Congrès archéol. de France*, 84 (1921), 67–89; Châtelain, *Donjons romans*, pp. 190–1, 200–1, 225, pls. xiv–xv, xl, xlii, xlviii.]

[6] [Dated *c.*1190: see P. Héliot and J. Vallery-Radot, 'Le donjon de la Roche-Guyon', *Mémoires de la Société historique et archéologique de l'arrondissement de Pontoise*, 58 (1962), 9–20.]

68. Niort, the *machicoulis*. [August 1908]

69. Chalusset

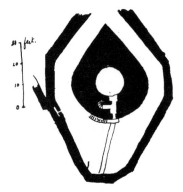

70. La Roche Guyon. Donjon
(Viollet-le-Duc [,*Dictionnaire
raisonné*, v, fig. 22])

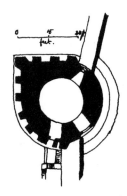

71. Château Gaillard. The donjon
(Viollet-le-Duc, *Dictionnaire*
[*raisonné*, v, fig. 31])

It had been recognised from the beginning that the undefended angles of Norman keeps were so many weak points, and that the remedy used by the Normans, shallow buttresses at each side of the corner, was inadequate. The pilasters become gradually more pronounced in Normandy; a semi-circular fillet runs up the centre as at Loches (Fig. 72);[7] then at Montbazon (Figs. 72, 73)[8] large semi-circular buttresses are applied to the corners at each side, and one in the centre of the face. Finally the type works itself out at Niort (Figs. 72, 74) in Poitou, where the two angle buttresses are combined and carried to the top of the tower as a circular tourelle. These keeps at Niort are exceedingly valuable. They were built early, apparently about 1180, and in spite of their shape are true Norman keeps. The tourelles are such only in name: in reality they are all solid, except one which from the first-floor level contains the staircase (Fig. 74a). The little buttresses on the face are also solid, to the top floor, and the entrance is 24 feet above the ground. It has a portcullis however.[9]

There are parallels to be found of the shape of Niort, but none of the peculiar arrangement whereby two identical donjons stand side by side, apparently without any very elaborate connecting works (Fig. 74). The small building that now fills the gap is comparatively modern: that it is not a copy of the original is proved by the fact that it stultifies the great *machicoulis* thrown over between the buttresses above. At the same time there must have been some link between the two towers: to leave each in isolation would be ridiculous.

The base of the donjon at Provins (Fig. 75) is probably of the twelfth century. In shape it is rectangular, but the corners have been chamfered off to make room for half-round buttresses of the same projection as the side walls. The upper part of the donjon has been so rebuilt as to be worthless: the doorways though are probably of the original design, and it is interesting to note the generosity of the architect, in providing four entrances for a tower 46 feet square. They are on the first-floor

[7] ['Loches . . . was splendid, a huge Norman keep in excellent preservation, and its corner buttresses with little colonettes as per plan. There is also a church with very fine Romanesque west door and narthex' (*Home Letters*, pp. 77–8, 8 Aug. 1908). See J. Vallery-Radot, 'Loches', *Congrès archéol. de France*, 106 (1948), 111–25; J.-F. Finò, *Forteresses de la France médiévale: construction—attaque—défense* (Paris, 1970), 399–403; Châtelain, *Donjons romans*, pp. 156–7, pls. IX, XXXII.]

[8] [G. M. d'Espinay, 'Montbazon', *Congrès archéol. de France*, 26 (1869), 209–11; Châtelain, *Donjons romans*, pp. 154–5, pls. IX, XXXI.]

[9] ['Magnificent: nothing could possibly have been more opportune or more interesting for my thesis. The castle is composed of two Norman keeps, each square, and quite ordinary inside, but outside each has a tower at each corner, and a little turret in between. The corner towers are quite solid: in fact as I learnt at Loches, they are only buttresses' (*Home Letters*, pp. 76–7). See also J. Bily-Brossard, *Le Château de Niort et son donjon: Notice historique et archéologique* (Niort, 1958); Finò, *Forteresses*, p. 171, fig. 45; Héliot, *Bull. archéol. du Comité*, 5 (1969), 169, 184, figs. 26–8; Châtelain, *Donjons romans*, pp. 178–80, pls. XII, XXXVI.]

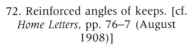

72. Reinforced angles of keeps. [cf. *Home Letters*, pp. 76–7 (August 1908)]

73. [Montbazon, the donjon]

247. NIORT (Deux-Sèvres) — Le Donjon.

74. [Niort, the donjon]

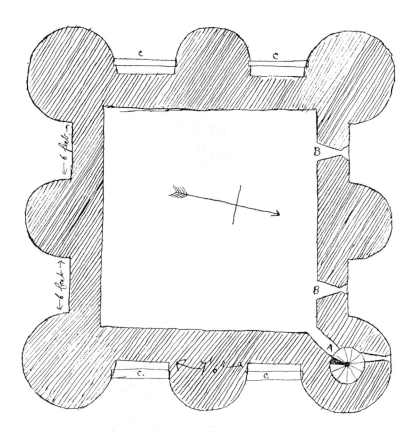

74*a*. The west donjon at Niort (19 August 1908). The centre tourelles, being semi-circular, project less than is here shown. A is the STAIR, leading up to the first and second floor. It ceases 10 ft. above ground level. B.B. are loops, splayed in and out. C.C. are the *machicoulis*, let into the circumference of the towers

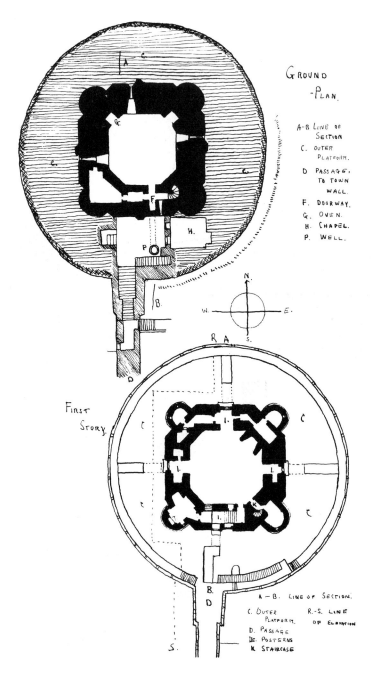

75*a*. [The keep at Provins. July 1908]

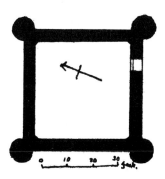

76. River tower near Tripoli. [August 1909]

level.[10] With Provins may be compared a little tower in Syria (Fig. 76), one of the links in the chain of defences that bound together Tripoli and its seaport (El Mina). It is almost the only attempt at variation on the square tower plan still surviving in the East: one can only conclude that the Latins were unfavourable to fancy designs.[11]

The keep of Étampes (Fig. 77) is perhaps the most astonishing production of the late twelfth century.[12] The square tower had been found wanting, and so the imagination of its architect conceived the idea of a quatrefoil tower which should be equally commodious, and a little less helpless before an attack. He kept the massive base of the Norman keep, and the entrance high in the air, but above that the shallow projection of the leaves was turned to account most ingeniously inside and out. When the tower stood complete (Fig. 78) with its hoardage of oak, not far short of 100 feet in height, it must have been no mean fortress: though it was commanded rather badly from the hillside at the back. With Niort and Provins, it will show the extraordinary life and vigour in the military architecture of the later twelfth century, an age which had outgrown the keep and was casting about for something more efficient to fulfil its purpose.

The best solution of the problem (which was not discovered till after polygonal towers had been tried and found wanting) proved to be the circular keep, standing isolated and self-contained in a moat of its own. Philip Augustus built multitudes of these, and Richard of England a few: finally Enguerrand de Coucy set the seal of his approval upon it (Fig. 78a).[13] One of Richard's keeps, at Montreuil Bonnin (Fig. 79),

[10] ['This has a most puzzling xii cent. keep, and remains of town walls. I was in and around them for hours, and came to the conclusion that the architect was making experiments when he built them. On the walls is a square tower that turned inside out and cut in half: the keep would have been a[l]most incapable of defence, and yet in spirit it is half a century ahead of its time. It ranks with Château Gaillard in importance for my thesis' (*Home Letters*, p. 61). The donjon dates from the second half of the 12th century: see P. Héliot and P. Rousseau, 'L'âge des donjons d'Étampes et de Provins', *Bulletin de la Société nationale des antiquaires de France* (1967), 289–308; Finò, *Forteresses*, pp. 416–20. On the town walls, see J. Mesqui, *Provins: La fortification d'une ville au Moyen Âge* (Bibliothèque de la Société française d'archéologie: Paris and Geneva, 1979).]

[11] [The towers defending the harbour at Tripoli are more likely to date to the Mamluk period: see H. Salamé-Sarkis, *Contribution à l'histoire de Tripoli et de sa région à l'époque des croisades* (Bibl. archéol. et hist., 106; Paris, 1980), 38 n. 1, map 8; Müller-Wiener, *Castles*, p. 42; P. D. A. Harvey, 'Drawings of Ports in the Levant, *c.*1600', *British Museum Quarterly*, 24. 3–4 (1961), 75–6, pl. xxiv.]

[12] [A date between 1130 and 1150 now seems more probable: see Héliot and Rousseau, *Bull. de la Soc. nat. des antiqs. de France* (1967), 289–98; Héliot, *Bull. archéol. du Comité*, 5 (1969), 166–8, figs. 24–5; cf. E. Lefèvre-Pontalis, 'Étampes', *Congrès archéol. de France*, 82 (1919), 3–49; Finò, *Forteresses*, pp. 356–7.]

[13] [Dated *c.*1230; see p. 39 n. 17. In July 1908, Lawrence wrote of Coucy, 'This was better even than Pierrefonds:— for one thing, it is xiii cent.—another its keep is 200 odd feet high . . . and there are splendid remains of four other towers, a great hall with two tiers of cellars beneath, and domestic buildings:— besides the town has almost complete walls around it' (*Home Letters*, p. 61).]

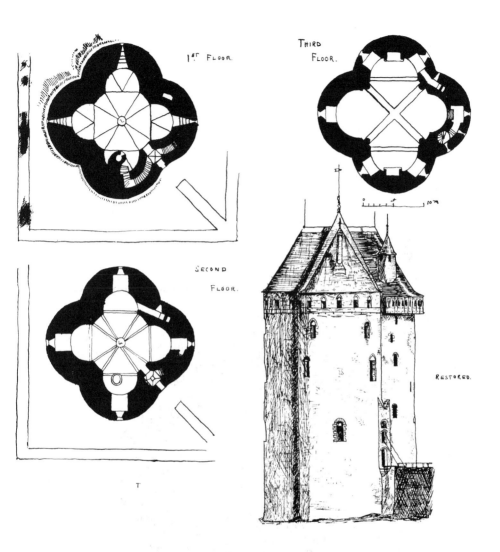

77–8. Keep of Étampes. [Viollet-le-Duc, *Dictionnaire raisonné*, v, figs. 15–17, 20]

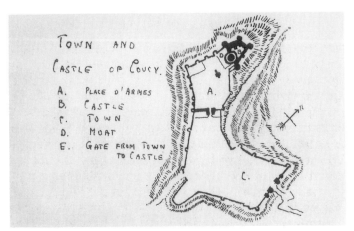

78a. Town and castle of Coucy. [Viollet-le-Duc, *Dictionnaire raisonné*, i, fig. 20]

A. Place d'armes   B. Castle

C. Town   D. Moat

E. Gate from town to castle

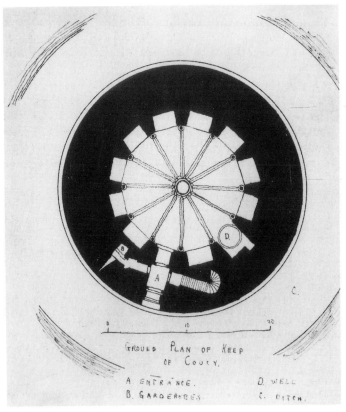

78b. Ground-plan of keep of Coucy. [Viollet le-Duc, fig. 41]

A. Entrance   C. Ditch

B. Garderobes   D. Well

near Poitiers, is almost certainly earlier than his Crusade.[14] The entrance is 30 feet in the air, and the staircase is straight instead of winding. The contemporary outworks and ditch show that Richard did not despise such helps to perfection. A nearer example of an eccentric circular keep will be found in Conisborough in Yorkshire.[15]

The shell-keep alternatives to the large square tower are wonderfully numerous, but that at Gisors is so very well known and so typical, that there is no need to give any description of them. It was built by Henry II (Fig. 80) but since his time the large door has been pierced in the wall, and the ungainly turret crushed in between two of the buttresses of the octagonal tower: this octagon has itself been rebuilt from the first-floor level.[16] A similar instance of shell-keep and tower within, is at Boves near by.[17] The keep of Gisors is a small one, standing as it does on an artificial mound, but there is one at Pujols (Fig. 87), on the Dordogne near Bordeaux, which encircles the top of a natural hill, and is of very considerable size. There are no signs of outer works, or of inner defences.[18]

It is a mistake to limit the activities of the twelfth-century builders to donjon-towers and shell-keeps. They were well accustomed to putting up more complex fortifications. To assert that the conception of a concentric castle had to be learnt from Byzantium, and imported laboriously into Europe just in time for the building of Château Gaillard, is to fly in the face of all probability. The architect of the 'early pointed' period, who, from his own intelligence was performing in church-building wonders that have never been surpassed in any age or country, was probably capable of the calculation that two walls were stronger than one or three than two. Had he not been he could have

[14] [Lawrence visited the castle in Aug. 1908: 'From Niort I rode towards Poitiers, turning aside at Montreuil Bellay, which castle Richard I is supposed to have built. The doorway bears an inscription in, I think, Arabic characters:— it has never been translated' (*Home Letters*, p. 77). The inscription, in Latin characters, shows that the castle existed by 1235. It now seems more probable that it was built by Philip Augustus after his conquest of Poitou in 1204. See R. Crozet, 'Le château de Montreuil-Bonnin', *Bulletin de la Société des antiquaires de l'ouest*, 4s 2 (1953), 507–14; cf. J. Thirion, 'Chronique', *Bulletin monumental*, 112 (1954), 286–7.]

[15] [The keep of the late 12th century is strengthened by 6 massive rectangular buttresses and has a battered base: see Renn, *Norman Castles*, pp. 155–7, figs. 24, 27, pl. x.]

[16] [Lawrence's plan was made on his second visit in 1908 (*Home Letters*, p. 60; cf. pp. 55, 109). The shell-keep and its octagonal tower are now considered to date from as early as 1096–7, in the reign of William II Rufus. Under Henry II (between 1161 and 1184) the tower was reroofed and its shell-keep provided with buttresses built of well-dressed ashlars. See Y. Bruand, 'Le château de Gisors: principales campagnes de construction', *Bulletin monumental*, 116 (1958), 243–65; cf. E. Pépin, *Gisors et la vallée de l'Epte* (Paris, 1963); Finò, *Forteresses*, pp. 372–6.]

[17] [See Châtelain, *Donjons romans*, pp. 100–1, pl. I.]

[18] It's full of Mairie and village school. [X] [The castle is not documented before the 13th century: see J. Gardelles, *Les châteaux du Moyen Âge dans la France du sud-ouest* (Bibl. de la Soc. fr. d'archéol., 3; Paris and Geneva, 1972), 201, figs. 111–19.]

79. Montreuil Bonnin

81. [Château Gaillard from the river]

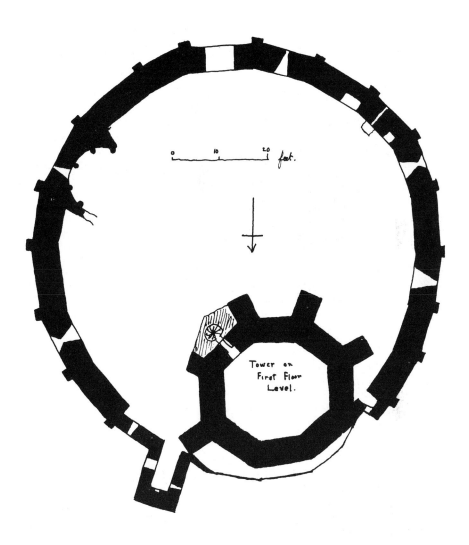

Tower on First Floor Level.

80. Gisors. The shell-keep. [August 1908]

looked at any earthwork or any Norman keep-and-bailey castle: he might even apply at the nearest monastery, and be told of the advantages of the triple wall praised by Lampert of Herschfeldt.[19] The monks might even lend him a copy of the Vegetius that Geoffrey Plantagenet studied with such profit in his sieges.[20]

Multiple castles as a matter of fact were built in Europe at all periods, becoming stronger and stronger with improved arts of attack. At Fréteval (Fig. 82), the twelfth-century architect ran a light stone walling round the top of existing double earthworks:[21] at Taillebourg he carved out ditches in soft rock, across a promontory running out into the marshes of the Charente.[22] At Chinon he did the same on a hillside.[23] At Hautefort (Fig. 83), near Périgueux, he cut one ditch across the neck of the ridge of hill, and led the entry along it through an outwork and then over it by a bridge.[24] There is really no necessity to assume that the architects of these and a hundred similar fortresses went to the trouble and expense of visiting the Holy Land, to learn from the Byzantines how to labour the obvious.

The kind of building erected within these ditches, one may gather from the castles of Carcassonne (Fig. 84) and Léhon (Fig. 85),[25] or the much-repaired 'tour du Moulin' at Chinon. There would be plain curtain walls with hoards, and lofty round towers at the angles; if the place was large enough, there would be salient towers half-round along the curtain as well. Very complicated defences were not required so early: they came in as they were required, for military architecture is less a series of miraculous improvements than a steady development along existing lines. The square keep was an exception without precedent, and without result: when it had been overpassed, the evolution of the form outlined in earthwork was resumed.

[19] [= Lambert of Hersfeld, died after 1077.]

[20] [Viollet-le-Duc, Dictionnaire raisonné, iii. 89 n. 1. The story is told in V. Mortet and P. Deschamps (eds.), Historia Gaufredi, Recueil des textes relatifs à l'histoire de l'architecture . . . en France au Moyen Âge, 2 vols. (Paris, 1911–19), ii. 81–2, no. 29 (1151).]

[21] [Described by Lawrence in Aug. 1908 as 'almost incredibly concentric but XII cent. all the same—a marvellous fortress' (Home Letters, p. 80). See A. de Dion, 'Le château de Fréteval', Bulletin monumental, 2 (1874), 205–15; id., 'Date de construction de la tour de Fréteval', Bulletin monumental, 6 (1878), 268–70; Finò, Forteresses, pp. 364–7; C. Leymarois, 'Fréteval', Archéologie médiévale, 12 (1982), 351–3.]

[22] [Gardelles, Châteaux . . . du sud-ouest, p. 225.]

[23] [Visited by Lawrence in Aug. 1908: 'Chinon is very fine indeed, (the French Windsor for its associations and place in history) but much destroyed. What there is, is of the xiii and xiv cents.' (Home Letters, p. 77). See R. Crozet, 'Chinon', Congrès archéol. de France, 106 (1948), 342–63; E. Pépin, Chinon (Paris, 1963).]

[24] [This arrangement, however, dates from the 17th century: see P. Vitry, 'Le château de Hautefort', Congrès archéol. de France, 90 (1927), 226–39.]

[25] [Lawrence had visited Léhon, near Dinan in Brittany, in Aug. 1906 (Home Letters, pp. 11–12).]

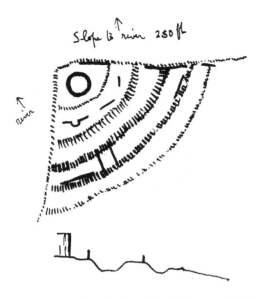

Slope to river 250 ft

82. Rough sketch of Frétéval. [August 1908]

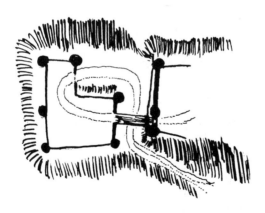

83. [Hautefort.] A very rough sketch: the walls are quite
imaginary, but the bridge and ditch are undoubted.
[August 1908]

84. The castle of Carcassonne (restored). (Viollet-le-Duc, [fig. 18])

84*a*. The castle of Carcassonne. [August 1908]

86. Montbrun

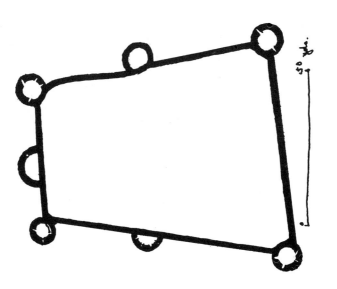

85. Léhon. [August 1906]

A few details must be cleared up as far as possible. In the vexed question of *machicoulis* one can come to no decision. The ordinary type are found apparently first in France, round three sides of a little tower in the castle of Montbrun, near Châlus in the Limousin (Fig. 86).[26] The tower is evidently Italian Romanesque (which is not surprising in twelfth-century Limoges) and so they are linked in a manner with the more finished *machicoulis* at Banias in Syria. The box type is found at Pujols (Fig. 87), in the hollow of the shallow pilasters at the angles of the polygonal shell. There are too many of them to have been insertions. These are probably earlier than the similar ones at Crac des Chevaliers; but marvellously inferior.[27] Further, they line the walls of Aigues-Mortes [Fig. 87a], built for Philip the Fair by an Italian contractor, at intervals of from 80 to 100 yards, which means that they are mostly sham:[28] and there are five on the Tour de l'Aubespin at Montbard in Burgundy, of the early fourteenth century. It is wisest not to propound theories on such evidence.

The large buttress-*machicoulis* are found at Niort, where they are a manifest addition of a later century (Fig. 68);[29] at Château Gaillard (Fig. 96);[30] round the church of Agde of the late twelfth century (Fig. 89);[31] on the Papal palace at Avignon (Fig. 88);[32] in the walls of Southampton;[33] and at Les Saintes Maries in the Camargue,[34] where

---

[26] [The tower is 12th century (after 1178), and the machicolations may not be much later: see C. de Beaumont, 'Tours du xıı⁽ᵉ⁾ siècle dans la région nontronnaise', *Congrès archéol. de France*, 79. 2 (1912), 345–64 (p. 360); Châtelain, *Donjons romans*, pp. 198–9, pls. xv, xlɪɪɪ.]

[27] [The *machicoulis* at Pujols also seem more likely to be 14th century: see Gardelles, *Châteaux . . . du sud-ouest*, p. 201.]

[28] [See *Inventaire général des monuments et des richesses artistiques de la France: Gard: Canton d'Aigues-Mortes* (Paris, 1973), i; cf. C. H. Bothamly, 'The Walled Town of Aigues-Mortes', *Archaeological Journal*, 73 (1916), 217–94; A. Fliche, 'Aigues-Mortes', *Congrès archéol. de France*, 108 (1950), 90–103; B. Sournia, 'Les fortifications d'Aigues-Mortes', *Congrès archéol. de France*, 134 (1976), 9–26; Finò, *Forteresses*, pp. 307–10.]

[29] [They probably date to around 1200; see n. 9.]

[30] [See below.]

[31] [Lawrence was evidently a little perplexed when he first saw this building in Aug. 1908: 'Agde, (superb fortified church a stumbling block to thesis)' (*Letters*, p. 58); 'a wonderful church . . . The building is all fortified, with a wonderful machicoulis all round. It has a front seat in my thesis' (*Home Letters*, p. 67). The church with machicolations dates from 1173 onwards. See J. Vallery-Radot, 'L'ancienne cathédrale Saint-Étienne d'Agde', *Congrès archéol. de France*, 108 (1950), 201–18.]

[32] [In the 14th century: see L.-H. Labande, *Le Palais des papes et les monuments d'Avignon au xıv⁽ᵉ⁾ siècle*, 2 vols. (Marseille and Aix-en-Provence, 1925); S. Gagnière, *Le Palais des papes d'Avignon* (Paris, 1974); Finò, *Forteresses*, pp. 319–23.]

[33] [Dated 1360 or 1378/9: B. H. StJ. O'Neil, 'Southampton Town Wall', in W. F. Grimes (ed.), *Aspects of Archaeology: Essays Presented to O. G. S. Crawford* (London, 1951), 243–57; P. Peberdy, *Historic Buildings of Southampton* (City Museums Publication, 5; Southampton, 1969), 35–7; D. F. Renn, 'The Southampton Arcade', *Medieval Archaeology*, 8 (1964), 226–8.]

[34] [Dated late 13th to early 14th century: see F. Benoit, 'Église des Saintes-Maries-de-la-Mer', *Bulletin monumental*, 95 (1936), 145–80; X. Barral i Altet, 'L'église fortifiée des Saintes-Maries-de-la-Mer', *Congrès archéol. de France*, 134 (1976), 240–66.]

87. Pujols

87a. [The walls of Aigues-Mortes. August 1908]

Hérault — 177 - AGDE
Cathédrale Saint-Étienne

AVIGNON — PALAIS DES PAPES
Tour des Anges vue de la cour extérieure

89. Agde. Saint-Étienne

88. Avignon. Palais des Papes

every other arch rests on a console.[35] This is a most illuminating list—
if it proves anything, it is that numerically Europe has the advantage
over the East: also that they were not the property of one school of
architecture or of one country.

Of other Byzantine features, the drawbridge pier is to be seen at
Tonquédec in Brittany, Coucy (thirteenth century) and at Chepstow,[36]
while square towers are only used on very rare occasions. There are a
few in the walls of Provins, a medley of every shape of tower conceiv-
able: round, pointed, square, and re-entrant towers jostle one another
in 300 yards of wall. The architect, whoever he was, was trying
experiments.[37]

Crussol, in the Rhône valley opposite Valence, is the only castle of
France presenting many Greek features. The gate (Fig. 90) is quite
Byzantine, and any curtain-towers there may be, are square (Fig. 91):
but the walls are so thin that a *chemin de ronde* had to be carried on the
butt-ends of the rests for the hoards, and the whole place was evidently
trivial.[38] The donjon was square, and reasonably solid: the other walls
were only to enclose a village. The place does not resemble in the least
anything European.[39]

There is therefore no room in all this for any borrowings from
Constantinople or the Templars in French architecture down to
Château Gaillard.[40] This unfortunate place is always quoted as an
example of the influence of the Crusades on medieval castle-building;
the opportunity of strengthening the statement possible in Richard's
visit to the East is too good to be missed. On the other hand quite
certainly there is nothing like Château Gaillard in the East. 'Un des plus
biaus chastiaus du monde, et des plus forz' as Guillaume Guiart
describes it,[41] it is nevertheless (or therefore) northern French in
design, and north French in execution. Richard undoubtedly devised it

[35] As in the cathedral (12th century) at Le Puy. [But the machicolations date from one or
two centuries later: see N. Thiollier, 'Guide archéologique du Congrès du Puy en 1904',
Congrès archéol. de France, 71 (1904), 3–91 (pp. 20–2); Finò, *Forteresses*, pp. 220–1, fig. 50.]

[36] [At Tonquédec the drawbridge appears to be 15th century (see p. 39 n. 17). On Coucy (a
pencilled addition to the text), see also p. 39 n. 17. Lawrence's description of Chepstow Castle,
made during a visit in Apr. 1907, is unfortunately only partly preserved (*Home Letters*, pp. 50–
1). The bridge-pier dates from the early 16th century: see J. C. Perks, *Chepstow Castle, Gwent*
(HMSO; 2nd edn. London, 1967, repr. 1975); Renn, *Norman Castles*, pp. 140–2.]

[37] [See n. 10.]

[38] It was a Byzantine habit to corbel out the *chemin de ronde*. [X]

[39] [The castle is mostly 12th century.]

[40] [Fortified by Richard Coeur de Lion between 1196 and 1198: M. Dieulafoy, 'Le Château-
Gaillard et l'architecture militaire au xiiie siècle', *Mémoires de l'Académie des inscriptions et belles
lettres*, 36. 1 (1898), 325–86; P. Héliot, 'Le Château-Gaillard et les forteresses des xiie et xiiie
siècles en Europe occidentale', *Château Gaillard: Études de castellologie européenne*, 1 (1964), 53–
75; L. Coutil, *Le Château-Gaillard*, 5th edn. (Les Andelys, 1928); Finò, *Forteresses*, pp. 336–40;
cf. Viollet-le-Duc, *Dictionnaire raisonné*, iii. 83–102, v. 69–72.]

[41] [Viollet-le-Duc, *Dictionnaire raisonné*, i. 89 n.]

90. Crussol. [July 1908]

90a. Crussol, Rhône valley. [July 1908]

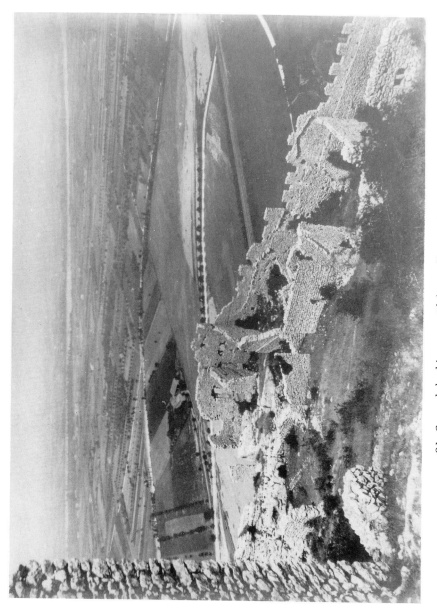

91. Crussol, looking over Rhône valley. [July 1908]

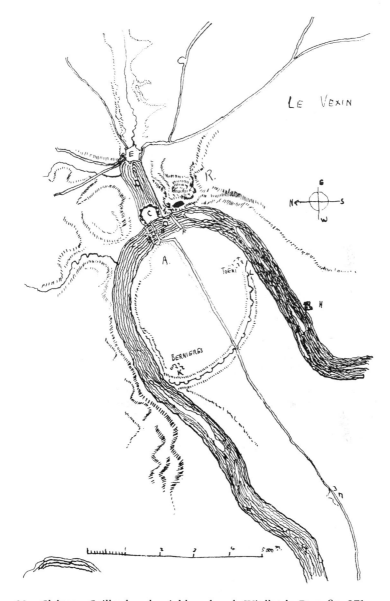

91a. Château-Gaillard and neighbourhood. [Viollet-le-Duc, fig. 27]

A. Head of the peninsula
B. Small island
C. Petit-Andely
D. Lake
E. Grand-Andely

F. Stockade
H. Bout-avant
K–L. Rampart of circumvallation
R. Plateau of Philip Augustus

himself: all authorities agree upon that, and throughout it shows a unity of purpose that could only have been secured by a consummate master of war, absolutely uncontrolled. Its plan (Fig. 92) shows an outwork with round towers, of quite ordinary character, behind a deep rock-cut ditch, and the outwork is cut off from the castle itself by another ditch, less deep. The outer ward has round towers everywhere except at one angle, where attack was absolutely impossible: most of the rest is destroyed, so that no gate can be identified. Within the outer ward, in the manner of Hautefort, is a deep moat, with the ribbed walls (Figs. 93, 94) of the inner ward rising sharply up from its edge. These ribbed walls have never yet been found anywhere else on earth. Viollet-le-Duc describes them most effectively, but cannot find a parallel. Semi-circular buttresses are common enough, and one finds a suggestion of Richard's plan in St Remi at Rheims;[42] or the Cathedral at Albi;[43] or in the now destroyed donjon at Condé-sur-Noireau described by De Caumont.[44] Probably Richard invented the idea: certainly no one copied it, so it cannot have met with approval. Inside all these walls is the donjon (Fig. 95), a small round tower with a spur (Fig. 71), and crowned with buttress-*machicoulis*. It was entered on the first floor, but is too small to stand a siege (Fig. 96).[45] Richard of course made it wonderfully strong, with its enormous talus (Fig. 95), and massive walling: but his garrison were not able to make use of it against Philip Augustus. The day of very small castles had gone by.

One detail of Château Gaillard, the horizontal striping of the walls by alternate courses of light and dark stone, is thought to be Byzantine on the analogy of some towers in the enceinte of Constantinople. Unfortunately no account of Richard's visit to that city has been preserved.

Château Gaillard is no exotic growth, but a development of the multiple castle of the style of Taillebourg and Hautefort, in the hands of an engineer of genius. There is no evidence that Richard borrowed anything, great or small, from any fortress which he saw in the Holy Land: it is not likely that he would do so, since he would find better examples of everything in that south of France which he knew so well. There is not a trace of anything Byzantine in the ordinary French castle, or in any English one: while there are evident signs that all that

---

[42] [L. Demaison, 'Église Saint-Remi', *Congrès archéol. de France*, 78. 1 (1911), 57–105 (pp. 89–90, fig. opp. p. 88). The church was partially destroyed during the First World War.]

[43] ['Albi: the cathedral looks like a blanc-mange mould: hideous but enormous and most imposing' (*Letters*, p. 58 (Aug. 1908); cf. *Home Letters*, p. 69). It was built between 1282 and 1383/97. See J. Laran, *La Cathédrale d'Albi* (Paris, 1944); E. Mâle, *La Cathédrale d'Albi* (Paris, 1950).]

[44] Not destroyed only hidden by a villain. I saw it in 1910. [X] [See p. 23 n. 50.]

[45] In Viollet-le-Duc's restoration [p. 91, fig. 31 = *Dictionnaire raisonné*, iii. 70, fig. 30], the figure of a man climbing the stair is about 20 ins. high.

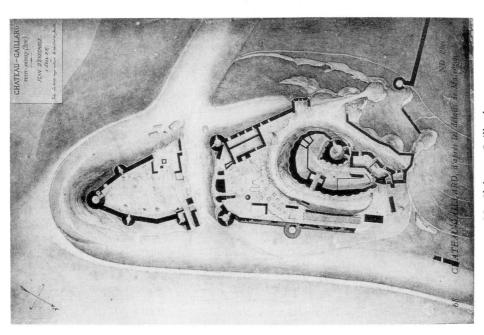

92a. Cellars of Château Gaillard. 1908

92. Château Gaillard

92*b*. [Cellars of Château Gaillard. ? August 1907]

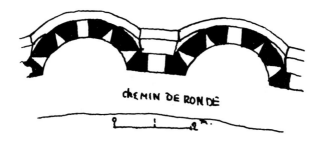

CHEMIN DE RONDE

93. Château Gaillard. *Chemin de ronde*

94, Château Gaillard, inner ward. [August 1907]

95. The donjon, Château Gaillard. [August 1907]

96. The donjon, Château Gaillard. [August 1907]

was good in Crusading architecture hailed from France or Italy. A summing up of the whole matter would be the statement that 'the Crusading architects were for many years copyists of the Western builders'.

APPENDIX I
EXTRACTS FROM A PRELIMINARY DRAFT OF
CRUSADER CASTLES (1909–1910)

Among the papers of T. E. Lawrence deposited in the Bodleian Library in Oxford there exists a bundle of notes (Res. C. 52, envelope 8) which evidently represents a preliminary draft for the Thesis which later became *Crusader Castles*. The extracts printed here are no more than fragments of continuous text followed by notes for the remaining portion. These were both to be developed further in the final version. Despite its tentative and provisional nature, the preliminary draft is of value, first because it begins with a remarkably clear statement of the Thesis itself, unencumbered by later elaboration; secondly, because it contains a number of descriptive details and observations that were omitted in the subsequent reworkings. It also provides some insights into the author's working method.

[II]

The classical view of the interrelation of East and West in the twelfth and thirteenth centuries in military architecture is that the first Crusaders took with them to Syria so primitive a style of castle-building that they abandoned it almost at once, to follow Byzantine models: in other words that the original conception of the Syrian fortresses is to be found rather in the walls of Constantinople, Nicaea or Antioch, than in Sicily or Normandy: and further that these new principles were carried to France by those returning, so that 'the Western builders were for many years timid copyists of the crusading architects'[1] and Château Gaillard itself is a 'typical example of a Syrian fortress, without parallel in the West for another quarter of a century'.[2]

I would suggest that the exact contrary is the case. That the Crusaders took with them to Syria the Norman keep as it stood (with only the modifications made necessary by the absence of heavy timber); that they failed altogether to improve upon it during the first century of

[1] Oman, *Art of War*, p. 532.

[2] [W. H.] Hutton, [*Philip Augustus* (Foreign Statesmen; London, 1896). Hutton's description of Château Gaillard, used extensively by Lawrence, includes such phrases as 'no castle in all Gaul could compare with Château Gaillard for the combination of natural and scientific defences' (p. 72, cf. pp. 61–2, 70–9); but the quotation cited by Lawrence is not to be found in it.]

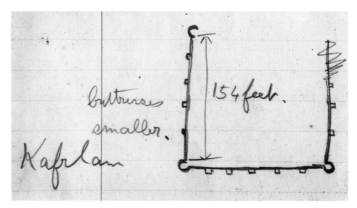

97. [Kafr Lām. Plan of rectangular early Muslim fort, with external buttresses and solid cylindrical corner-turrets. Lawrence thanks C. H. C. Pirie-Gordon for providing him with the original of this plan (p. 315), but he does not appear to have used it in his Thesis. For more details of the site, see Benvenisti, *Crusaders*, pp. 281–2, 329–31]

98. [Acre. The harbour and Tower of the Flies, from the north. Mount Carmel in the distance. 1909]

their occupation; that their borrowing from Byzantium meant if anything a loss of efficiency in their fortresses, since in many respects the Greeks had fallen behind the West in their architecture; and that it is only when powerful monarchs, such as Richard I, and Louis IX visited Syria, that any real advance in its castle-building can be seen. The few really great castles of Syria, Crac des Chevaliers (Kala'at el Hosn), Margat (Markab), and Banias, seem to me all of thirteenth-century construction, all later than the visit of Louis IX who brought with him to Syria his favourite architect, Eudes,[3] to reproduce (as at Sidon) his own style of fortress on Syrian soil.

The classical view is based on the assumption that outer wards and foreworks, numerous and strong curtain-towers are borrowings from Byzantine work, and that the provision of flanking (or covering) fire for exposed points was another derivation from the East, with the general idea of what is called the 'concentric' castle. (*There never were any concentric castles* later than the earthworks: Castle Rising, Old Sarum, Hembury, Maiden Castle: the shape would have been hopelessly unsatisfactory: the word can never be taken to mean more than multiple defences on the weakest side.) I think that all these ideas, with the details necessary to establish them are to be found earlier in France than in Syria: and some of them not at all in Byzantine building. Since Justinian the art of military engineering among the Greeks had made little progress . . .

[V]

The Crusaders occupied Syria piecemeal, and in their local divisions generally denoting race-differences, one would expect to find different schools of architecture predominant. Thus Edessa used Byzantine models entirely: indeed generally, the Franks built in that province nothing at all: the Greeks before them had fortified sufficiently, and they were content. Antioch built little that is recognisable. Tripoli, as the chief centre of the activities of the Templars and Hospitallers, contains enormous fortresses. The buildings of the Kingdom itself have suffered most from later destructive agencies: the huge castle of Kerak in the Desert, of which there is a plan and description in Rey's book,[4] has unfortunately never been visited by a medievalist.[5] Rey himself only describes at second hand: and the existence of signs of activity of

---

[3] [Eudes of Châteauroux, Bishop of Tusculum, was the Papal Legate, not an architect.]

[4] [Rey, pp. 3, 17, 131–5, 273–4, pl. XIV.]

[5] The unthinking action of some Bedawin in tearing up the Hedjaz railway prevented my going there in 1909.

Beibars makes doubtful the elaborate deductions drawn by him from its plan as to the state of Frankish fortification in 1140. Indeed except in rare instances mere dates, on documentary evidence, of the building of strongholds, are utterly valueless: sometimes the actual remains show that Byzantine remains were already on the site: at other times the castle has visibly been rebuilt, perhaps by a later generation of Crusaders, or by Beibars (who generously took under his aegis, by cutting inscriptions, the towers of his predecessors), or even as in extreme cases, at Sidon and Belfort, by Arabs of the sixteenth and seventeenth century. Quite generally, however, Arab architectural influence on Frank architects may be entirely discounted: they appear never to have had any fortresses of importance, until Beibars learnt how to build from his Greek and Latin enemies: and put up the outer wall of Crac des Chevaliers, and the citadels of Aleppo, and Damascus. Before him one may find the masterpieces of the Arabs in the trifling constructions at Kadmus and Masyad.

For a certain period after their arrival the Franks contented themselves with erecting slightly modified[6] square keeps: which form belonged exclusively to them: no Byzantine building presents anything analogous. Square keeps are to be found in Sahyun, Kalaat Yahmur, Safita, Batrun, Jebeil, and Kalaat esh Sh'kif. Something of the kind seems to have existed also at Caesarea and Safed and Beaufort: and a keep of square shape, but with turrets at the two ends (very doubtful) is marked by Rey in his plan of Tartus. The ruins on the spot, much encumbered with houses, and harems impossible to enter, appear to point rather to a square tower abutting on two curtain-towers of the seaward wall. It is however quite beyond verification.

Some of the keeps are sufficiently original in plan to make necessary closer study. That at Jebeil is said (quite wrongly) by Rey to be probably the earliest remaining of Crusading buildings in Syria. It is oblong in shape, 78 feet × 55, and probably stood nearly 100 feet in height (the battlements are ruined etc.). The doorway on the ground floor has a portcullis; but the new feature in the building is the parapet, rising some 14 feet above the level of the uppermost floor in the building, and containing two stages of defence after the manner of the Byzantine buildings of the East.

The keep of Batrun is mostly destroyed: and that of Kalaat esh Sh'kif likewise: they were smaller than the others.

The keep of Safita is in some respects the most important. It is a huge tower 100 feet × 60, and standing, complete today to the battlements,

---

[6] The modifications are such as would reasonably be expected taking into consideration the absence of wood as building material. The keeps become smaller and are vaulted with stone: and the entrance on the first floor usually disappears.

well over 100 feet above the crest of the hill, down-sloping sharply on all sides to a ring wall, of Templar construction, 100 feet lower in level. The entrance, a small door, closed by a plain hinged gate, leads direct into the village church, defended slightly by one loophole high above the altar, and two on each side. They were however so narrow, and at such a height above the ground, as to be evidently unfitted for defence. Indeed usually, loopholes are meant less for firing through, than for light: their arches are seldom made high enough (or wide enough) for longbow or crossbow and the field covered by an archer within the tower firing a long arrow is extremely slight. In the New College loopholes an arc of fire of only 21 yards at a distance of 100 is possible. Those at Safita are evidently only to light the church or chapel of the fortress; and do not do it very well at that. (The stair up is defensible: absurdly.) Above the chapel is a hall, well lighted, with low loops: and up a stair from this the roof of the tower with a single row of loops, and merlons with recessed sockets for a protecting shutter. These were the usual means of defence in Syria until *machicoulis* were introduced. This parapet is possibly the oldest preserved unrestored of medieval French builders.

This keep, from its position, was almost secure against siege engines but its smallness made it insufficient for the garrison: Safita was one of the chief fiefs of Tripoli. Accordingly a wall was built a little lower down the hill on the east at some time unknown.[7] And when the Templars recovered the fief in the thirteenth century they put round the hill a huge wall, polygonal, in their pseudo-Byzantine style, to enclose the part of the village that was defensible. This wall is visibly of Templar work: just as the central donjon is evidently in the mid-twelfth century.[8] The simplicity of its parapet is noteworthy.

The keep at Sahyun is the most massive of Crusader keeps. It measures about 90 feet square, and has walls over 20 feet in thickness. In the ground floor is the entrance, a narrow gateway closed only by a door, without portcullis: indeed of the keeps only Jebeil, probably the latest, has the portcullis. The gateway gives on a flight of stairs, leading in the thickness of the wall (and undefended) to the upper floor. The lower is lit by two narrow loopholes, so high up that the room is almost dark. In the centre is a huge pier, supporting a lofty, but rather flat-vaulted ceiling. The upper floor is lit by five loopholes of reasonable size, fully defensible, and has latrines: the staircase to the roof, again in the wall (newel stairs are not common in Palestine) leads on to a

---

[7] This had been grubbed away a few months before I was there to provide houses for the increasing family of the present governor of the district.

[8] [The first draft reads 'early in the 12th century'. This was changed to 'late', before the final compromise was reached.]

terrace, surrounded with high parapet. This is a transition before that of Jebeil: leading from Safita.

A third most important keep is that of Kalaʿat Yahmur (Chastel Rouge). The history of this is unknown, except that it was handed over to the Templars, at some uncertain time.[9] Within a peculiar rectangular outwork stands a stone keep, 54 feet by 73, quite perfect, entered on the first floor, by a species of terrace between it and the western wall. Habitations beneath, and all round make it impossible to investigate. The first-floor room is vaulted on a central pier: and by the loopholes there is undoubtedly a cellar beneath: the platform above has a parapet and battlements of two stages of defence, which commands fairly well the rectangle of covering walls. Two machicolated towers appear certainly to be rebuilding of the upper parts of towers. The gateway is quite exceptional, and possibly of late date. The complicated defences outside have (but for their mason-marks) a peculiarly Byzantine appearance.[10] Of course it is possible that the square keep was enclosed by the Templars in the thirteenth century by a wall of their usual character: only they appear never to have used *machicoulis*: and the entrance to the keep on the first floor shows that outbuildings of some sort were understood in the primitive plan. The castle stands in the open plain and is almost entirely without ditches (except for a faint suggestion on the north and south): and it will be seen that the enclosing walls are placed extremely close to the central tower. The gateway is comparable with that on the north wall of Tripoli [Ṭarṭūs].

The tower at Caesarea is too destroyed for more than an understanding that it was large, and rectangular: but that at Safed was standing a century ago fairly complete: and the remains still existing, show a species of small tower (of the dimensions of Oxford Castle tower) on the highest part of the castle hill. It was put up apparently in 1140: and repaired or rebuilt half a century later by the Templars. They then remodelled the mound.

At Belvoir (Coquet or Kaubab el Howa) was a square enceinte, with square towers at the angles and on the curtain, ranging within a wide rock-hewn moat: and on the cliff side, overlooking 2,000 feet down the valley of the Jordan, a large square keep, of the style of Jebail. It was built in the second half of the twelfth century, and is of the same quality as Hunin.

It will be seen from the foregoing that the Latins never allowed their square keeps to stand alone: though the outworks that are now standing are usually not those built by the builders of the keeps. These keeps were all built by the feudal lords of the fiefs: before the coming of

---

[9] [In fact it was given to the Hospitallers in 1177.]

[10] [The outer defences of the gate no longer survive.]

the great Military Orders, and they were put up to compare with those that they had left behind them in Provence or Sicily. There is visible in them no admiration of Byzantine principles of defence, unless it be in the arrangements of the parapets in some examples. In the walls around them however we find very distinct traces derived from the Greeks, and amounting even in cases to a deliberate borrowing of an already fortified site.

In its early stages this adaptation is easy to trace. In the Principalities of Edessa and Antioch the Crusaders found magnificent fortresses ready to their hand. They besieged Antioch: and found it so difficult to capture that they never troubled to alter its defences. Edessa they left as they found it: just as they left the people their Byzantine code of laws, and employed Greek officials in their government. The castle is still standing at Edessa with all its essential features: except the great φρουρά [φρούριον], the Persian tower, a species of Byzantine keep, which lay probably at the western extremity of the castle, from the ruins of the huge hexagonal tower still standing there. Perhaps from this, or a similar tower or the other small one close by, the French architect of Loches, of Chalusset, and the gates of Provins, drew his inspiration. It presented a beaked face toward the outer country, so that a mangonel stone struck only a glancing blow, and a ram became almost ineffective. Another feature found very occasionally in Europe, the pier left in the moat to hold up the centre of the drawbridge, appears here, and also, to a far more sensational extent, in Sahyun. Other castles in the district of Edessa, Tell Bashar, which is entirely destroyed with the exception of a square bastion on the mound, Biredjik, of later than Crusader building, and Rum Kala'at with its two portcullis, solitary witness of Latin ownership, show Byzantine influence, in so far as they may be traced.

In the Principality of Antioch the few castles that have survived to modern times, are almost equally Byzantine. Harim (Harenc) is entered up its huge rock-cut moat, until a ramp ascends to the gate on the western face: this gate may have been remade in Arab times: at least it is remarkably like a piece of Byzantine building at Kizil Hissar (Ramsay and Bell . . .).[11] Antioch itself is all Byzantine: and Bagras was held by Greeks and Armenians, rather than by Latins in the twelfth century. Other castles are unidentified somewhere in the riot of hills that extends southwards and westwards from the valley of the Orontes: so that it is not till Sahyun, inland from Latakia, is reached that any considerable building is discovered. Sahyun has the square keep described on page 127, but in addition it is one of the strongest buildings

----

[11] [*The Thousand and One Churches* (London, 1909), 279–80.]

of the Byzantine time that has come down to us in Syria. The plan[12] shows the gigantic moat, rock-cut, 90 feet deep, and in parts as much wide: with the single slender pinnacle standing up in the centre to support the drawbridge. The ditch is cut through from side to side of the narrow promontory, with impassably steep sides, on which the castle is built. There is a round tower at the south-east angle, and square towers along each side, with postern entrances in C and D (plan). These gates are only some 4 feet wide, and are closed by a door. The towers themselves are cut off from the *chemin de ronde* of the curtain in the typical Byzantine manner, and that lining the great ditch has a second tier of defences, also Byzantine in style. The castle proper ends in F–G, but the lower end of the promontory is defended by a low breastwork (which the fall of the cliff made impregnable) enclosing enough ground for a considerable colony: this spur has a separate entrance (below which the valley has been bridged) and is isolated entirely from the upper fort, not by a ditch, but by a sudden terrace, some 50 feet in height, of bare rock.

There are quarter-round towers on each side of the drawbridge gateway.

At Jebeil the keep is surrounded by a wall, rectangular, with square angle-towers of Byzantine style. The masonry is of quite different quality to that of the keep, and approximates more nearly to that of the town wall. The difference in date between these three constructions need not however be very great: for King Amaury in 11 . . .[13] built at Darum on the borders of the desert of Egypt a fortress described by William of Tyre in terms that are identical with what could be used of many Greek castles in northern Africa. The Byzantine influence was not long in getting a foothold in Latin Syria.

So far the distinction of styles has been fairly easy. We find Crusaders erecting their own familiar European strongholds, so far as possible within the walls of Byzantine buildings: this was usually done on quite a modest scale: the ordinary eastern feudatory was not sufficiently rich to put up anything exceptionally large: but the gradual increase of the power of the two great Military Orders,[14] which took the form of annexing to themselves all the frontier fortresses in the Kingdom, led to a sudden revival of military architecture, and the development of new ideas and styles in fortification. Nine-tenths of the castles in Syria were

---

[12] I have corrected some of the errors made by Rey in his pl. xxvi [cf. Fig. 22]. I did not recognise until I checked it with my notes how peculiarly inaccurate it is. The section of the entrance gateway is unrecognisable: and the north side (as well preserved as the south) he has left out entirely. His stay was either extremely short, or his notes were not made on the spot.

[13] [By 1170.]

[14] The German Order came later—in the early 13th century—and erected Montfort, a typical German Rhine-castle, between Tyre and ʿAkka.

added to or rebuilt in the last twenty years of the twelfth century: and to add to the difficulty, the rivalry of the two Orders led them to adopt different styles for their buildings: the Templars, always suspected of Eastern heresies and sympathies, took the castles of the Byzantine builders for their models, and simplified their ideas. The Hospitallers drew their inspiration from the flourishing school of military architects in France, and with their help put up the gigantic fortresses of Markab, and Kala'at el Hosn and the rest, which proved the last word in Syrian castle-building, and were copied in the Armenian buildings, and in the buildings of the Venetians and their allies in Greek waters. The two schools have entirely different ideals, and principles: and the two classes of buildings stand entirely apart, without a link or compromise between them: and the history of the sieges of the different types shows that in defensive power the castles of the Hospital were the better. Markab and Crac were considered by the Arabs more formidable than Athlit or Safed.[15]

The Templar Castles

It seems easiest to take these, the Byzantine models, first, since they affected Europe little: and their rivals in Syria not the least. The Templars in building such places as Athlit rejected all the carefully arranged flanking fire, and covering works, and lines of multiple defence which were being thought out meanwhile in Europe. At Athlit they relied on the one line of defence—an enormously thick wall, with two scarcely projecting square towers upon it: these were the keeps, the master-towers of the fortress, and instead of being cunningly arranged where they would be least approachable, they are placed across the danger line, to bear the full force of the attack. They had of course no *machicoulis*, and their projection was not sufficient to shield the curtain wall. Their strength lay in their brute strength and solidity, and the barrier to mining represented by a deep, sea-level trench, excavated before them.

. . .

[VI]

4. The corresponding European progress of the twelfth century:
a. i. The shell-keep: at Gisors, and Pujols. Enlarged and transformed at Léhon and Carcassonne: in the latter place on a Roman site.

[15] [There is no evidence for this assertion: 'Atlīt was never taken by assault, and Ṣafad fell to Baybars only through the treachery of one of its defenders.]

    ii. The variations on the square keep, Conisborough, Chalusset, Provins, Niort (via Loches and Montbazon), Étampes.

    iii. Hoarding. Crussol (permanent), Loudun, Oxford.

b. The concentric (or multiple) castle:

Castle Rising	where the earthworks antedate the keep
Les Baux	a city wall makes a second line
Polignac	a natural rock suggests a tower
Crussol	a village is built round the castle: and enclosed
Niort	two keeps side by side
Gisors	where a tower is added to the shell-keep
Châlus	two castles defend a third in the valley
Montreuil Bonnin	Richard digs a ditch and wall round a circular keep
Frétéval	a round keep is placed inside prehistoric earthworks which are improved by walls
Taillebourg	where three ditches are cut across a promontory
Hautefort	which adds an outwork.

c. Eastern comparisons:

1. No keep variations in Syria.

2. No shell-keeps in Syria: and nothing to approach Léhon or Carcassonne.

3. The elaborate entrance is found only at Hautefort. In Carcassonne, and Pierrefonds[16] and Tonquédec (of late thirteenth century) are Eastern coincidences: not copies. Mondoubleau[17] is almost the only gateway comparable with Syria.

4. Beaked towers with square, half-round, and re-entrant appear in Provins. These are visibly experiments. The square tower did not get a footing in Europe till the fourteenth century.

   Portcullises were not used in the East: except at Tartus and Crac: and the talus was not used in the West except at Château Gaillard. (Newel staircases in West usually. Straight in Montreuil Bonnin.)

   Of *machicoulis* the buttress type is found at Crac: but in the West at Agde and Avignon (!), Southampton, and Château Gaillard. The example at Crac is visibly Italian or Provençal.

   Corbelled *machicoulis* are found at Les Saintes Maries (with buttresses), and at Montbrun. In the East best at Banias (Italian). Boxed *machicoulis* at Crac, Aleppo etc., at Pujols, Aigues-Mortes, and Montbard. In Europe they often served as latrines.

---

[16] [Lawrence visited Pierrefonds in July 1908 (*Home Letters*, pp. 60–1).]

[17] [Lawrence visited the keep at Mondoubleau and planned the doorway in Aug. 1908 (*Home Letters*, pp. 79–80).]

5. Château Gaillard has therefore an outwork like Hautefort: and middle and inner wards like Taillebourg.

Its inner ward is unlike anything in East or West. The nearest approaches being Étamps, Niort, Albi, Saint Remi at Rheims, and Condé-sur-Noireau. Richard probably invented it himself: no one imitated.

The keep has a 'spur', like Chalusset: and buttress-*machicoulis*, like Agde: and a unique 'talus'.

The outer walls had hoarding only: and this mainly brought about the collapse of the building.

The decoration of the outer wall, in horizontal stripes, is Norman; and so are the window openings.

Château Gaillard is therefore the work of an originally minded engineer, who had borrowed precious little from his sojourn in the East.

APPENDIX II
THE STRATEGIC SITING OF CRUSADER CASTLES
(1911)

Lawrence's brief account of the defensive strategy of the Crusader states in the Levant was written in response to a request for help from Leonard Green, who was to give some lectures on the Crusades in Oxford. Lawrence's reply, in a letter written from Jubayl, is undated but was received by Green on 14 January 1911. It was first published in full in *Letters*, pp. 93–7.

What I felt most myself in Syria, put shortly, was the extreme difficulty of the country. Esdraelon, and the plain in which Baalbek lies are the only flat places in it. The coast road is often only 50 yards wide between hills and sea, and these hills you cannot walk or ride over, because they are strewn over with large and small boulders, without an inch of cultivated soil: also numberless small 'wadies' (torrent-beds) deep and precipitous: not to be crossed without a huge scramble. In one day's march, from Lake Huleh to Safed, one ascends and descends 16,000 feet in hills and valleys, often 1,500 feet deep and only 200 yards or so across, and in all the way only a single track path on which one can ride without fear of smashing horse's legs. Make a point that for heavy-armed horse operations in such country are impossible: they can march in single file, but cannot scout, or prepare against surprise: the battle-field of Hattin (near Tiberias) is like a dried lava-flow, or the photo-graph (only in rock) of the pack-ice of the Arctic seas. . . . Even when there are no mountains or rivers there will be hills and valleys enough, with rock-stretches, to make an impassable tract. You will never, without seeing it, conceive of the difficulty of the country. On the main road from Antioch to Aleppo my escort walked with their horses (after Harim) for nearly four hours: and for a Syrian to fare afoot is much against the grain.

The next point is the rivers: the Jordan is hardly passable except at three points: just below Lake Huleh (Gisr Benat Yakub, today: Castle Jacob or le Chastellet in the Crusade authorities) a bridge and ford. Another ford just below lake Tiberias (near Semakh)[1] and one more (very difficult) near Jericho. The first two were available for or against the Damascenes. From Lake Huleh northwards is a swamp and the river Litani, until the hills get steep enough and high enough to be

---

[1] [This is presumably the Pont de Senabra (near Sinn al-Nābra), at the southern end of the lake.]

impassable: and then (very quickly) comes the Orontes, which is nearly always impassable (from Riblah downwards) to Esh Shogr, on the road from Latakia (Laodicea) to Aleppo. There is a ford there, and after a bridge near Antioch (the Iron Bridge). Above Antioch came a large lake, and then very hilly ground from the Kara Su to Alexandretta. So you see west Syria is pretty well defended. In the early days of the Latin Kingdom they held all this, and as well pushed across the Euphrates (via Harim, Tell Bashar, and Biredjik) to Urfa (Edessa). This was a sort of outpost, which kept apart the Arabs of Mesopotamia and the hills to the north (Kurdistan) from the Arabs of east Syria (Aleppo, Homs, Damascus) and the Arabians. While the Crusaders held Edessa, which is a tremendous fortress (of Justinian's), they were unassailable except through the Damascus gap, and one opposite Homs (Emesa) extending to Tripoli. This last gap (which I forgot to mention before) is a nearly sea-level pass between Lebanon and the Nozairiyeh hills. It was defended by three tremendous Crusader castles (Crac des Chevaliers, in Arabic Kalaat el Hosn; Safita, Chastel Blanc in the French authorities; and Aarka, just above Tripoli). These three places made this gap tolerably harmless, except from Arab raids: these were continuous: but only did temporary harm: still they neutralised the force of the county of Tripoli from 1140 onwards. The Damascus gaps were also blocked: the northern one by Banias beyond Huleh, by Hunin, above the lake, by Safed, and by Chastel Jacob, just beside the ford, which last castle only existed a few months. Stevenson in his book points out its importance rather more cheerfully than usual with him.[2] The southern ford, below Lake Tiberias, was defended by the town of Tiberias, by Kaukab el Hawa (one of the Belvoirs, just above Beisan) and by an outpost, el Husn, occasionally held beyond Jordan: it is not marked on any map.[3] The Jericho ford was never very important. There were some little Crusader castles on its Syrian side.

The whole history of the Crusades was a struggle for the possession of these castles: the Arabs were never dispossessed of Aleppo, or of Hamah, or of Homs, or of Damascus, and so they had all possible routes open to them: they had unlimited resources to draw upon, as soon as the Mesopotamians had recovered Urfa (Edessa), which the Crusaders

---

[2] [W. B. Stevenson, *The Crusaders in the East* (Cambridge, 1907).]

[3] [Qalʿat al-Ḥiṣn, sited on a promontory overlooking the Sea of Galilee, is identified with Hippos, a city of the Decapolis. The walls appear to be Roman and Byzantine, with no trace of any medieval refortification: see G. Schumacher, *The Jaulan* (Palestine Exploration Fund; London, 1888), 194–206; and C. Epstein, 'Hippos (Sussita)', in M. Avi-Yonah and E. Stern (eds.), *Encyclopedia of Archaeological Excavations in the Holy Land* (Oxford, 1976), ii. 521–3. Although al-Ḥiṣn does not appear to have been occupied by the Crusaders, Habīs Jaldak, a cave-fortress in the Yarmuq Gorge, and Qalʿat Bardawīl in the southern Jaulan certainly were.]

could not hold on account of its isolated position (Euphrates 10 feet deep, 150 yards wide, very rapid, and often flooded, much difficult hill thence to Seraj, and even nearer Edessa) and the shiftiness of the Greek Armenian population, who were allies, at times, but fighting men not at all: more harm than good usually. The native population of Syria very much sympathised with the Arabs, except the Maronite Christians and the Armenians; and news travels amazingly in the East: so that the Latins were more often surprised than not. Any counterstroke in the nature of ambush against the Arabs was impossible, since half their people were spies. Then in the hills above Safita lived the Assassins (Haschishīn) sometimes at war with the Arabs, more often confeder-ate, and linked with them by an Orontes-bridge at Shaizar (Kalaat Seidjar). They could not attack Tripoli because of the 'gap' castles (see before) and Markab (Margat) a huge fortress north of Tartus (another stronghold): but they could and did hold Antioch in check from the south, while the Aleppines pressed on the Iron Bridge, and the Greeks and Arabs attacked by Alexandretta and Beilan and Bagras (the two last big castles).[4] So Antioch could only just hold its own, and the Tripoli castles, and when Damascus (Noureddin) joined with the north, and added Egypt, Syria was ringed round. The battle of Hattin was lost in an attempt to relieve Tiberias, the second of the great 'gap' fortresses. Banias (the first) was lost about 1150.[5] For most of the occupation the Latin sphere of influence was limited to their castles: the peasantry paid them taxes, and wished for the Mohammedans to come: and come they very often did, to plunder such Christian villages as were left. So far as one can see they spared the Mohammedans. Latin Syria lived on its fleets.

This is horribly condensed, and much platitudinous: if only I could tell you what you wanted to know: and there is no book written by a man who has been out, except Baedeker! Do emphasise the importance of the fortresses, which are all marked in a map in my 'Thesis'. Salaams,

E.L.

[4] [Here Lawrence is in error. Baghras was the single major castle defending the Baylan (or Belen) Pass through the Amanus Mountains between Antioch and Alexandretta on the coast. See A. W. Lawrence, 'The Castle of Baghras', in T. S. R. Boase (ed.), *The Cilician Kingdom of Armenia* (Edinburgh and London, 1978), 34–83; cf. R. W. Edwards, 'Bağras and Armenian Cilicia: A Reassessment'. *Revue des études arméniennes*, NS 17 (1983), 415–55, pls. LX–LXXXII.]
[5] [1164.]

APPENDIX III
DESCRIPTION OF FIVE CASTLES IN THE COUNTY OF EDESSA (1911)

In July 1911, following the conclusion of the British Museum's excavation at Carchemish with which he was then involved, T. E. Lawrence spent some four weeks on a walking tour north-east of the Euphrates. Starting from Tall Aḥmar, he visited the castles of Urfa (Edessa), Ḥarrān, Birijik (al-Bīra), Rum Kalaat and Tall Bāshir. Details of the journey were kept in a pencilled diary, portions of which are printed below.

The diary was first published in full in 1937 under the title, *The Diary of T. E. Lawrence* (London and New York), and was reprinted in a more exact version with some of the original photographs in *Oriental Assembly* (edited by A. W. Lawrence, London, 1939). As A. W. Lawrence notes in his introduction to the latter edition, 'Some of the photographs mentioned in the Diary cannot be traced, while others, not mentioned in the text, were obviously taken on this journey.' The extracts published here are from the version printed in *Oriental Assembly*. They are accompanied by a selection of the photographs that did not appear in that edition, but were evidently taken in 1911. They may easily be distinguished from the photographs which Lawrence took in 1909, for whereas the latter were taken on a plate size of 5½ by 4½ inches, these are all 4 by 3 inches.

EDESSA[1] (URFA)

Friday[, *14 July*]:

. . . Urfa about midday . . . Took room in great khan: then went out about 4 p.m. to photograph castle. Took it from the due west [*OA*, pl. III] showing the double gates and the line of walls from the πυργοκάστελλον to the extreme end. . . .

Saturday[, *15 July*]:

Up late (about 6 a.m.) and went out to the castle. Photographed the castle at the south-east angle: where the moat turns, and above which is one of the very few Crusader walls in existence here [*OA*, pl. IV reversed]. It is patched in front (to right) with Arab wall, but is very fine. A wide-angle photo. Then measured the east side of the moat, and

---

[1] [See p. 26 n. 3 above.]

photographed the east half of the south side, by wide angle from the bottom of the moat [Fig. 99]. This makes complete my photographs of the moat, all but the north side. Then measured this east half of the south side, and went and had some bread. . . . After lunch went back to the castle and measured till 5 p.m. Decided the north side moat did not deserve a photo: average depth of moat about 40 feet. Greatest present depth 60 feet, but much filled in. Crusade work is to be found in patches in the entrance gateways, at the south-east angle-tower, and in a piece of the north wall. On coming down took a photo of the castle from a little street that runs north-east [*OA*, pl. v reversed]. This view of the north-east angle of the castle and the back of the gate-towers looked pretty on account of the amount of green about. In the khan I found the chief of police and a follower, who remonstrated with me for going about alone. 'Boys might throw stones' etc. He insists on a zaptieh [escort] tomorrow. . . .

Sunday[, *16 July*]:

Up late (8 a.m.) and had a great wash: found police waiting for me all round the khan; went up the castle with one little man. He complained of the heat, so I sat him under an arch with some snow and a bowl of water and tobacco, and he was happy.

Measured the interior etc. A fresh morning with a cool west breeze. Took a photo of the interior of the castle from the tall beaked tower at the west end [*OA*, pl. vi]: breeze rather troublesome, but could not get the tripod up: climb rather difficult. The angle-tower is altogether Arabic. Later on photographed the great gateway (also Arabic) from the top of a tower [Fig. 100]. Decided that almost everything in the place was Arabic except the moat, some straight pieces of wall, and the south-west angle-tower: with the two Roman pillars. . . .

Monday, 17 July:

Up about 4 a.m., but was a long time getting on the road . . . Bought a metallik of bread, and went over to the castle. Town wall 9–10 feet thick. About 6 started for Harran. . . .

HARRĀN[2]

Country everywhere as flat as possible: only huge tells about every two miles . . . The tower of Harran Cathedral was in sight for four hours: all

---

[2] [See S. Lloyd and W. Brice, 'Harran', *Anatolian Studies*, 1 (1951), 77–111; D. S. Rice, 'Medieval Ḥarrān: Studies on its Topography and Monuments, I', *Anatolian Studies*, 2 (1952), 36–84.]

100. [Edessa (Urfa). The great gateway to the castle. November 1911]

99. [Edessa (Urfa). The south moat of the castle, looking west from the south-east corner. Compare *Oriental Assembly*, pl. IV, which, however, is printed back to front. November 1911]

elongated by the mirage, it becked and bobbed in the most fantastic way, now shivering from top to bottom, now bowing to right or left, now a deep curtsey forward. . . . Stopped outside Harran walls for a short rest, then climbed through a gap into the town. The main part of the village lay to the south-east of the old site, around the castle. . . . I found the Sheikh in the castle, which he has made his house. There was a huge stone-vaulted polygonal tower, with deep embrasures and an earth floor. . . .

Tuesday, 18 July:

Up by daybreak, and round the outside of the castle. The inside I had explored with the Sheikh the afternoon before. . . . The castle built at several periods: part of it quite late; none apparently pre-Arab: mostly of rusticated blocks: there was no ornament anywhere. Huge polygonal towers flank the outer wall, and there is a sort of keep, of smooth stone, with shallow buttress-towers at the corners: inside, this is vaulted on two square pillars in one room, others have plain barrel vaults. The castle has had a moat round it: perhaps a wet one. It has been a big strong place, but not over-interesting. The vaulting though is good. Then went in and drank coffee (four cups) with the Sheikh and his men: . . . Later walked over to the mosque, and looked for Miss Bell's column-capitals. Took a photo [*OA*, pl. VII] of a lion bas-relief in basalt— 5 feet 2 inches long, 3 feet 6 inches high, 1 foot thick: broken in two pieces: rude work. Muzzle broken: lying just outside the east angle of the town wall. A boy behind. Was found on the surface of the ground. Then took a photo of the south front of the castle, not of the whole of it, but of the eastern half: this showed one small polygonal tower, and a line of walls, with the 'donjon' in the centre [*OA*, pl. VIII]. Then walked round and took a photo of the great broken tower [Fig 101]. Looking into it one could see the floors and the central pier, and the rest of the works of the place. This tower stands on the west side of the castle, defending one side looking towards the town. The former (south) side looks towards the open desert. Then went across again to the great mosque: could not turn over the other great capital, and found the little ones much damaged underneath: that is, the two I partially cleared. Not very interesting, these little ones.

Then started out seriously to take the Sheikh. Had taken him on horseback with his brother before the south front of the castle [*OA*, pl. VIII], and now took him with a friend of his before the tower-room [*OA*, pls. IX–X]. . . .

Then we fed, about 9 a.m. . . . Worked at the castle after lunch: measuring etc. Then walked across to Rebekah's well. I came in past it yesterday, resting near it half an hour . . . The well is down steps and

101. [Harrān. West corner-tower of the castle. November 1911]

102. [Birijik (al-Bīra). The northern half of the castle from the north-east.
November 1911]

very deep, cold, clean water. There are camel troughs near it, possibl[y] those that Eleazar used, for such things do not soon wear out. Good water. Drank again today. They call it Bir Yakub, and are very proud of it. It is the only well outside the walls. I saw also the Aleppo gate, a poor Arab thing, more ornamental than defensive: in fact the walls of Harran are slight defences: it is certainly not fortified for a siege, with its long thin curtains, and shallow towers, all square angled. The castle is the only fortress. There was a moat, probably wet, all round the town, and between it and the castle. There are no surface signs of pre-Byzantine occupation. . . . I went over to the castle again, and decided it was all fairly late: post-Saladin at least, possibly post-Crusade. No more photographs needed. The great broken tower is about 60 feet high. . . .

BIRIJIK CASTLE[3] (AL-BĪRA)

Saturday, 22 July:
. . . lay up in the khan reading and sleeping till 4 p.m. Then went out to the top of the hill, and photographed the town walls etc. from the south. The castle would be behind this hill a little to the left. Then went down into the valley and up hill again. Took the north half of the castle from the north-east, in the shade against the sun [Fig. 102]: and the south half of the castle (both landward side[s]) also from the north-east,'a little further on than the one before, and under the same disadvantage of light [*OA*, pl. XIII]. This finished my films loaded. . . .

RUM KALAAT[4] (QALʿAT AL-RŪM)

Monday, 24 July:
. . . Reached Rum Kalaat about 10 a.m. The place enormous, a town rather than a fortress. At first came visible part of a huge rock-moat, which cut off the peninsula on the south (land) side; then the scarp of the Euphrates wall, about 60–90 feet of rock-cutting. I had then to walk up the side-stream valley to the gate of the place, before I could cross it on wide stepping-stones: a broad swift stream, shallow. There was once a bridge. Walked round the far side of the little valley, half-way up. Took a general view, wide angle, showing the side stream, the hills and the Euphrates [Fig. 103]. Another, a little farther on, another (ordinary

---

[3] [See above, p. 39.]

[4] [See above, p. 39 n. 18. In a letter to his mother from Jarāblus, dated 29 July 1911, Lawrence wrote, 'the castle of Rum Kalaʿat yielded some new points, mostly Arab: it had a most enormous moat, a perfectly appalling thing: . . . It cut off a mountain from a mountain along a col like the coupée at Sark' (*Letters*, pp. 117–18; *Home Letters*, p. 176, cf. pls. between pp. 256 and 257).]

103. [Rum Kalaat, from the north-west, showing the Mezman Su flowing into the Euphrates. November 1911]

104. [Rum Kalaat. North-east valley scarp. November 1911]

lens) of the north-east valley scarp, in shadow mostly [Fig. 104]. This
may be a little fogged. Then went on to the mouth of the valley and took
one of the Euphrates front. This has a little domed building like a weli in
the foreground [Fig. 105]. Felt sleepy, so went to cave, and slept till
2 p.m. Got up then, and (i) telephotographed the box-*machicoulis* of
the north-west angle with a magnification of 13, and a stop of 22°:
exposure 12 secs., on normal of 1/50 nom. f.16. This was a large-scale
photo of three *machicoulis* [*OA*, pl. XIV]. Also (ii) telephotographed at
$3\frac{1}{2}$ mags. LP all the range of *machicoulis* (*c*.16) at f.11 and an exposure
of $\frac{1}{2}$ second [Fig. 106]. Both taken from the shade, and (i) with hood.
These *machicoulis* very remarkable. More about them later. Then went
down into castle (down and up!). Through five gates, all double and
protected by towers, one monolith, into the outer-court. This in shape
of narrow ledge, running north and south, gate to south. Builders of
this place not satisfied with 90 foot wall and scarp, absolutely perpendicular; but put a rock-moat outside as well: moat once wide and deep;
now all stuff of the walls and a graveyard have filled it up. The castle
as a whole occupies the narrow point of a peninsula, a rocky ridge,
pointing due north and south. This is surrounded on the east by the
Euphrates, on the west by the little river Mezman Su, and on the north
by the same: the south end is thus the only part not precipitous. The
crest of the ridge must be between 3 and 400 feet high. This is at the
south end, the highest, but not so high as the rock beyond the castle to
north and south, from both of which it was overlooked, though at a fair
distance off. The walls on east and west run about half-way up the ridge
and from inside them the rocks and ruins pile up, very steeply, to the
central pinnacles. The highest point of all is very elaborately carved,
and may have been a palace, or a church. The local[s] say a minaret,
which is probable, afterwards, but all the ornament is not Arab. The
building in the north corner of the ridge-crest is a mosque, with paved
court about it. Between this and the 'palace' all is destroyed, except
substructions and deep cellars cut in the rock. The view is limited, but
tremendous. The present village rests across the stream, on the north
bank where it turns east and west and extends into the Euphrates.
There are poplar trees, and the noise of water. The ridge at the south
end is about 30 feet broad at the top. This is cut down 90 feet to a path
about 8 feet wide, like the razor in Westmorland. The moat is about
60 feet wide. I took a photo of it from the Euphrates side, on a point of
the castle about 30 feet above the edge of the razor [*OA*, pl. XVI]. This is
not satisfactory, but gives the river flowing at the bottom very nicely.
After this I left the castle (6 p.m.) very tired, but a most glorious place,
and crossed the Mezman Su again by the crazy stepping-stones: the
hardest I have ever walked over. . . .

105. [Rum Kalaat. The east front, from the north-east. November 1911]

106. [Rum Kalaat. Row of box-*machicoulis* on the north-west front. November 1911]

108. [Rum Kalaat. Rock-cut main gate, with its applied groin-vaulting. November 1911]

107. [Rum Kalaat. Rock-moat from the west side, looking across the Mezman Su. November 1911]

Tuesday, 25 July:

Up at 3.45 (dawn) and had a wash in the stream . . . Wrote up this for a time and then stayed to eat, for there is no house but the cave-dwellers between this village and my night-stopping-place. We had burghul and bread together. Then I went along the over-river west-side path, till I could photograph the rock-moat [Fig. 107], and returned across the passage perilous, the stepping-stones that I know fairly well by now, to the castle. . . . Took a photo of the inside of the monolith tower, showing the applied vault [Fig. 108]. Tower about 17 feet wide inside between the inner jambs: the third gate counting from outside. The fourth gate, though also a monolith, I did not think worth a photo, since it is only a single arch. The fifth is a very fine Arab double-arched gate [*OA*, pl. XVII]. All this entrance-masonry is Arab, and very good.

The first two gates have *machicoulis* over them. Altogether one of the strongest and cleverest entrances in existence. The manner in which the roadway is made to double on itself, so that it may be more easily under control, and the right-angled turns at most of the gateways are especially clever. There are no trap *machicoulis* in the floors, so far as these are preserved, and there were no portcullises. The box-*machicoulis* of the north-west angle are very small inside: only one tiny loophole in front, none to the side. A photograph was tried (under lighting difficulties) of the inside of the vault with which they communicate. There is another vault below this, to serve more loopholes. A photograph (wide angled) was also tried of the razor, looking almost directly down upon it [*OA*, pl. XVIII]. The lens was not wide enough to include all the moat, so the lower part is cut off: There should be enough all the same to make it fairly intelligible. I made a few alterations in my Antinous [shutter-release]. Then I looked at the mosque on the north part of the ridge-crest. It is quite plain, with a date of 1236 [of the Hegira, AD 1820/21] on a side-door. It is probably very late. The whole place is full of Arab ruins. There is only the foundation of Byzantine stuff anywhere remaining. I left the castle about 9 a.m. by a postern door in a tower on the river side, and walked to Khalfati . . .

TURBESSEL[5] (TALL BĀSHIR)

Thursday, 27 July:

. . . Went on to Tell Bashar, where I stopped to glance at it, bigger than ever as it was, and turned off left at once for Tchiflik . . .

---

[5] [See above, p. 39 n. 13.]

INDEX OF PROPER NAMES

The names of modern countries, indicated in square brackets, are abbreviated as follows: Alg(eria), Eng(land), Eg(ypt), Fr(ance), Isr(ael), It(aly), Jor(dan), Leb(anon), Pal(estine: WB = West Bank, G = Gaza Strip), Sco(tland), Syr(ia), Tur(key), Wa(les). Alternative names or spellings are given in parentheses.